New Nanomaterials and their Electromagnetic Properties (NNEP 2021)

IOP Conference Series: Materials Science and Engineering
Volume 1198

Tomsk, Russia
4 – 6 April 2021

ISBN: 978-1-7138-4657-4
ISSN: 1757-8981

Printed from e-media with permission by:

Curran Associates, Inc.
57 Morehouse Lane
Red Hook, NY 12571

Some format issues inherent in the e-media version may also appear in this print version.

Institute of Physics
Dirac House, Temple Back
Bristol BS1 6BE UK

Phone: 44 1 17 929 7481
Fax: 44 1 17 920 0979

techtracking@iop.org

Additional copies of this publication are available from:

Curran Associates, Inc.
57 Morehouse Lane
Red Hook, NY 12571 USA
Phone: 845-758-0400
Fax: 845-758-2633
Email: curran@proceedings.com
Web: www.proceedings.com

TABLE OF CONTENTS

IOP Conf. Series: Materials Science and Engineering 1198 (2021) 011001 doi:10.1088/1757-899X/1198/1/011001

Fifth Russian-Belarusian Seminar "New Nanomaterials and their Electromagnetic Properties"

Abbreviated name of the seminar conference: NNEP-2021

The purpose of the seminar is to discuss the latest achievements in fundamental and applied science in the development, manufacture, application of new materials and devices based on them. The spectra of the electromagnetic characteristics of materials and the electromagnetic response of structures made of these materials, in a wide frequency range, including THz range, are also considered. Special attention is paid to modern measuring instruments and teaching methods for high school students and graduate students.

The first Russian-Belarusian International Seminar was held at the Belarusian State University in 2016 (Minsk), then at the Institute of Inorganic Chemistry SB RAS named after I. Nikolaeva A.V. in 2017 (Novosibirsk), at the Skolkovo Institute of Science and Technology in 2018 (Skolkovo), at the National Research Tomsk State University (TSU) in 2019 (Tomsk).

Notable speakers:
Dunaevsky G.E., Doctor of Technical Sciences, Professor, TSU, Tomsk.
Kuzhir P.P., Ph.D., Belarusian State University (BSU), Minsk, Institute of Photonics, University of Eastern Finland,
Maksimenko S.A., Doctor of Physical and Mathematical Sciences, Professor, BSU, Minsk,
Nasibulin A.G., Doctor of Technical Sciences, Professor, Skolkovo Institute of Science and Technology, Moscow,
Okotrub A.V., Doctor of Physical and Mathematical Sciences, Professor of Novosibirsk State University (NSU), Novosibirsk,
Fedorov G.E., Doctor of Physical and Mathematical Sciences, Professor of Moscow State Pedagogical University, Moscow Institute of Physics and Technology, Moscow,
Minin I.V., Doctor of Physical and Mathematical Sciences, Professor, Siberian State University of Geosystems and Technologies, Novosibirsk,
Kuznetsov V.L., Ph.D. Associate Professor, NSU, Institute of Coal of the Siberian Branch of the RAS, Novosibirsk,
Nomoev A.V., Doctor of Physical and Mathematical Sciences, Institute of Physical Material Science of the Siberian Branch of the RAS, Ulan-Ude, Russia,
Bashkuev U.B., Professor, Institute of Physical Material Science of the Siberian Branch of the RAS, Ulan-Ude, Russia,
Suslyaev V.I., Ph.D., Associate Professor of TSU, Tomsk.

Members of the Scientific and Organizing Committees:
Chairman Suslyaev V.I., Ph.D., Associate Professor, TSU, Tomsk,
Deputy Chairman Minin O.V., Doctor of Physical and Mathematical Sciences, Professor, Novosibirsk,
Scientific Secretary Dotsenko O.A., Candidate of Physical and Mathematical Sciences, Associate Professor, TSU, Tomsk.
Committee members:
Kochetkova T.D., Ph.D., associate professor of TSU,
Badin A.V., Ph.D., associate professor of TSU,
Zhukov A.A., Ph.D., associate professor of TSU,
Wagner D.V., Ph.D., associate Professor of TSU,
Korovin E.Yu., Ph.D., associate professor of TSU,
Kuleshov G.E., Ph.D., associate professor of TSU,
Pavlova A.A., post-graduate student of TSU,

IOP Conf. Series: Materials Science and Engineering 1198 (2021) 011001 doi:10.1088/1757-899X/1198/1/011001

Dorozhkin K.V., post-graduate student of TSU.

All materials were reviewed by the editors: Dr. U.L. Mishigdorzhiin from the Institute of Physical Material Science of the Siberian Branch of the RAS, Ulan-Ude, Russia and Prof. S.P. Bardakhanov from the Khristianovich Institute of Theoretical and Applied Mechanics, Siberian Branch of the Russian Academy of Sciences, Novosibirsk, Russia.

The conference consisted of plenary presentations for 30 minutes and session reports for 10 minutes. There were 35 oral reports. The organizers have provided 5 minutes for questions that can be asked in person or via the Internet.

Topics were:

• Synthesis and diagnostics of carbon nanostructures

• Hybrid materials

• Electrophysical characteristics of carbon nanostructures

• Application of THz and GHz radiation for characterization of nanomaterials

• Graphene electronics

• Magnetic nanomaterials

• Theoretical methods and approaches applied to nanoelectromagnetism

• Biomedical applications of new materials

• Automation of measurements

• Modern educational methods

The organizers and participants of NNEP-2021 noted the high scientific level of the presented reports.

IOP Conf. Series: Materials Science and Engineering 1198 (2021) 011002 doi:10.1088/1757-899X/1198/1/011002

Peer review declaration

All papers published in this volume of IOP Conference Series: Materials Science and Engineering have been peer reviewed through processes administered by the Editors. Reviews were conducted by expert referees to the professional and scientific standards expected of a proceedings journal published by IOP Publishing.

- **Type of peer review:** The examination was carried out by two experts. Since the number of all works is small - 15. All works were reviewed by two experts.
  When preparing for publications, we asked the authors to make two versions of articles in Russian and English language. Then the check was carried out first in Russian, then in English. The quality of the translation at the first stage of the check was checked by the Grammarly program - the site: https://www.grammarly.com/, at the second stage - it was assessed by a reviewer who has an excellent level of language English, Undrakh Mishigdorzhiin.

- **Conference submission management system:**
  The conference was managed by the organizing committee Chairman V.I.Suslyaev, Ph.D., associate professor, TSU, Tomsk, a member of the organizing committee A.V. Nomoev. , Doctor of Physics and Mathematics

- **Number of submissions received: 15**

- **Number of submissions sent for review: 15**

- **Number of submissions accepted: 15**

- **Acceptance Rate (Number of Submissions Accepted / Number of Submissions Received X 100): 100%**

- **Average number of reviews per paper: 2**

- **Total number of reviewers involved: 2**

- **Any additional info on review process:**

  The check was carried out by two experts. In the case of comments, errors, the need to add additional data to the peer-reviewed article, this was indicated to the authors of the article.
  They were corrected by the authors, re-checked by an expert for correcting comments, making changes and additions.

- **Contact person for queries:**
  Name : Andrey Nomoev
  Affiliation: Institute of Physical Materials Science, Siberian Branch of the Russian Academy of Sciences

IOP Conf. Series: Materials Science and Engineering 1198 (2021) 011002 doi:10.1088/1757-899X/1198/1/011002

Email : nomoevav@mail.ru

IOP Conf. Series: Materials Science and Engineering 1198 (2021) 012001 doi:10.1088/1757-899X/1198/1/012001

Low-frequency (10^5 Hz) shear modulus of nanosuspension

B B Badmaev, T S Dembelova, D N Makarova, S A Balzhinov and E D Vershinina

Institute of Physical Materials Science, Siberian Branch of the Russian Academy of Sciences, 670047, Ulan-Ude, Russia

E-mail: vershinina.eugenia@gmail.com

Abstract. The low-frequency shear wave propagation in a suspension of silica nanoparticles in a polyethylsiloxane liquid was studied. The shear wave length was measured on ultrasonic interferometer, and it is equal to 55 μm. The value of the tangent of the mechanical losses angle is determined to be 0.18. These parameters were used to calculate the shear modulus of the investigated colloidal suspension; its value is $0.15 \cdot 10^5$ Pa. The results obtained are in quite agreement with the data obtained by another way of the acoustic resonance method.

1. Introduction

Nanosuspension is a dispersed two-phase system consisting of a dispersion medium and a dispersed phase. Nanosuspension has special structural and mechanical properties, such as elasticity, viscosity, strength and ductility, which are manifested during shear deformation. The structure and rheological properties of the nanosuspension are mainly determined by the interaction of the particles of the dispersed phase with the molecules of the dispersed medium and with each other.

Studies of the rheological properties of nanosuspensions have been actively pursued over the past two decades. It was found that these properties depend on the size, shape, kind and concentration of nanoparticles [1]. However, there are still no unambiguous conclusions regarding the nature of these parameters of nanoparticles influence on the rheological properties of nanosuspensions [2, 3]. Therefore, experimental studies of the rheological properties of nanosuspensions by different methods are relevant. It is also of great practical importance in connection with the numerous areas of practical application.

In [4-6], the shear elasticity of SiO_2 nanosuspension in PES-2 at small thicknesses of the liquid layer was measured by the acoustic resonance method. However, the result obtained can be attributed to the special properties of the boundary layers of nanosuspensions. Therefore, a study of the possibility of a low-frequency shear wave propagation in a thick layer of nanosuspension would confirm that shear elasticity is its bulk property.

2. Method

In this work, the propagation of a low-frequency (10^5 Hz) shear wave in a colloidal suspension of nanoparticles of silica SiO_2 in a polyethylsiloxane liquid PES-2 is studied using an acoustic resonance method on an ultrasonic interferometer. The value of the shear modulus of the investigated nanosuspension from the measured values of the shear wave length and the tangent of the mechanical losses angle was calculated.

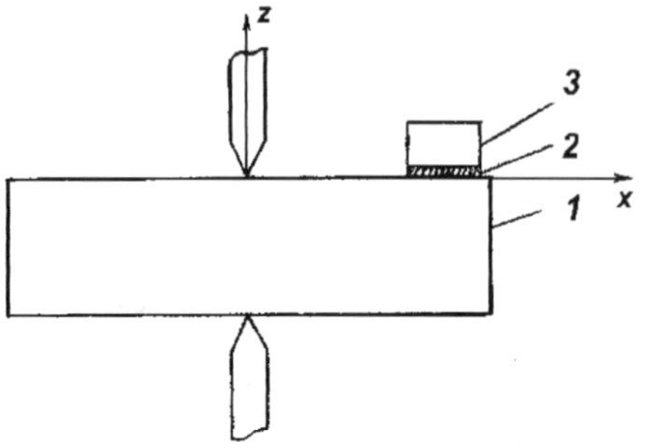

Figure 1. Piezoquartz with an additional bond: 1 – piezoquartz, 2 – liquid film, 3 –cover plate.

The essence of the acoustic resonance method is as follows. The horizontal surface of the piezoquartz crystal X-18.5° cut in the form of a rectangular bar with dimensions of 34.9x12x6 mm, weighing 6.82 g, contacts the liquid layer covered with a solid cover plate [7-9]. The cover plate is located at one end of the piezoelectric quartz, figure 1. When the piezoelectric quartz oscillates at a resonant frequency of 73.2 kHz, the liquid layer experiences dynamic shear deformations and a standing shear wave must be established in it. Depending on the thickness of the liquid layer, the resonance frequency of the piezoquartz and the width of the resonance curve change. This is the expression of the complex shift of the resonant frequency Δf^* [7-9]:

$$\Delta f^* = \frac{SG^*k^*}{4\pi^2 M f_0} \cdot \frac{1+\cos(2k^*H-\phi^*)}{\sin(2k^*H-\phi^*)}, \tag{1}$$

where $G^* = G' + iG''$ is the complex shear modulus of the liquid, S is the area of the base of the cover plate, $k^*=\beta - i\alpha$ is its complex wave number, H is the thickness of the liquid layer between the piezoquartz and the cover plate, ϕ^* is the complex phase shift that occurs when a viscoelastic wave is reflected from the liquid – cover plate interface, M is the mass of the piezoquartz, and f_0 is its resonant frequency.

Given that $\phi^* = 0$ the following expressions are obtained by dividing Δf^* into real $\Delta f'$ and imaginary $\Delta f''$ frequency shifts [7-9]:

$$\Delta f' = \frac{S}{4\pi^2 M f_0} \cdot \frac{(G'\beta + G''\alpha)\sin 2\beta H + (G'\alpha - G''\beta)\,\mathrm{sh}\,2\alpha H}{\mathrm{ch}\,2\alpha H - \cos 2\beta H}, \tag{2}$$

$$\Delta f'' = \frac{S}{4\pi^2 M f_0} \cdot \frac{(G''\beta - G'\alpha)\sin 2\beta H + (G''\alpha + G'\beta)\,\mathrm{sh}\,2\alpha H}{\mathrm{ch}\,2\alpha H - \cos 2\beta H}. \tag{3}$$

Using the ratio $(k^*)^2 = \frac{\omega^2 \rho}{G^*}$ it can be seen from these expressions, that the frequency shifts give damped oscillations with an increase in the thickness of the liquid layer. The maximum attenuation, which is observed in the opposite phase of the forward and reverse waves, can be used to determine the length of the shear wave λ. The first maximum attenuation will be observed when the thickness of the liquid layer is $\lambda/2$. We can obtain a calculated formula for the real shear modulus in the form by using the expression (3) [7-9]:

$$G' = \lambda^2 f_0^2 \rho \cos\theta \cos^2\frac{\theta}{2}, \tag{4}$$

where θ is the angle of mechanical losses, ρ is the density of the liquid under study. The value of tan θ is determined by the ratio of the distance ΔH between the first minimum and maximum of the real frequency shift to the length of the shear wave λ [7-9].

Thus G' and tan θ can be determined by measuring the shear wave length in a thick liquid layer.

A polyethylsiloxane liquid PES-2, which is a mixture of polymers of linear $(C_2H_5)_3Si$-O-$[Si(C_2H_5)_2O]_3$-$Si(C_2H_5)_3$ and cyclic $[(C_2H_5)_2SiO]_3$ structures, was used as a base liquid to obtain a nanosuspension. Table 1 shows the main properties of the polyethylsiloxane liquid PES-2 [10]. And a silicon dioxide nanosize powder SiO_2 with a size of 50 nm, trade mark Tarkosil T-0.5 were used as a dispersed phase.

Table 1. The main properties of the polyelilsiloxane liquid PES-2.

Molecular weight	Density at 20°C, kg/m^3	Viscosity, mm^2/s			Boiling point, °C (at 133 – 400 Pa)	Pour point, °C
		at -60 °C	at 20 °C	at 60 °C		
341	940	312	9	4	110 – 150	-110

3. Results and discussion

Another way of defining G^* follows from the analysis of expressions (2) and (3). If the thickness of the liquid layer H is much less than the shear wave length λ, i.e. $H << \lambda$, the expressions (2) and (3) are extremely simplified. For the real G' and imaginary G'' shear modulus, as well as for tan θ, the following formulas are obtained [7-9]:

$$G' = \frac{4\pi^2 M f_0 \Delta f' H}{S}, \quad G'' = \frac{4\pi^2 M f_0 \Delta f'' H}{S}, \quad \tan\theta = \frac{G''}{G'} = \frac{\Delta f''}{\Delta f'}. \tag{5}$$

It can be seen from these formulas, that the complex shift of the resonant frequency is proportional to the inverse of the thickness of the liquid layer. Thus, in this method of measuring G^*, it is sufficient to measure the thickness of the liquid layer H, and the real shift of the resonant frequency $\Delta f'$ and the imaginary shift $\Delta f''$ by the broadening of the resonant curve. In [7-9], this method was used to measure the complex shear modulus of a colloidal suspension of silica SiO_2 nanoparticles in a polyethylsiloxane liquid PES-2 at small thicknesses of the liquid layer. Linear dependences of the frequency shifts on the inverse thickness of the liquid layer were obtained. The values of the shear modulus and the tangent of the mechanical loss angle were calculated using formulas (5). The dependences of these parameters on the size of nanoinclusions and their concentration were investigated. Thus, for 0.5 wt.% with particles dimensions of 50 nm, the values of the shear modulus $G = 0.17 \cdot 10^5$ Pa and the tangent of the mechanical losses angle tan $\theta = 0.18$ were obtained [4].

A stable SiO_2/PES-2 nanosuspension was obtained as follows. After the introduction of silicon dioxide nanoparticles into the base liquid, the resulting dispersed system was thoroughly dispersed by ultrasound. This made it possible to destroy the agglomerates of nanoparticles as much as possible and to obtain their uniform distribution in the bulk of the dispersed medium [4-6]. The cover plate with the liquid layer was freely located at one end of the piezoelectric quartz when measuring the complex shear modulus of the nanosuspension at small thicknesses of the liquid layer, when $H << \lambda$. The thickness of the liquid layer was changed by displacing the liquid from under the cover plate. But when measuring G′ by the propagation of a shear wave, the cover plate was fixed rigidly and moved vertically by mechanical devices, which made it possible to regulate the thickness of the liquid layer over a wide range. Thus, an ultrasonic interferometer was implemented for low-frequency shear waves, described in detail in [7-9].

Figure 2 shows the experimental and theoretical dependences of the real $\Delta f'$ and imaginary $\Delta f''$ frequency shifts as a function of the thickness of the SiO_2/PES-2 liquid layer. Curve 1 refers to the real frequency shift, and curve 2 – to the imaginary one. As can be seen from the figure, there is a damped oscillation of frequency shifts. The shear wave length in a given nanosuspension can be determined from the attenuation maxima (curve 2), and it is equal to 55 μm. With respect to the $\Delta H/\lambda$ ratio, the value of the tangent of the mechanical losses angle is determined to be tan $\theta = 0.18$. The value of the shear modulus is calculated according to the formula (4), and it is equal to $G' = 0.15 \cdot 10^5$ Pa. The results of the G' measurement obtained by the two methods of the acoustic resonance method are quite consistent with each other.

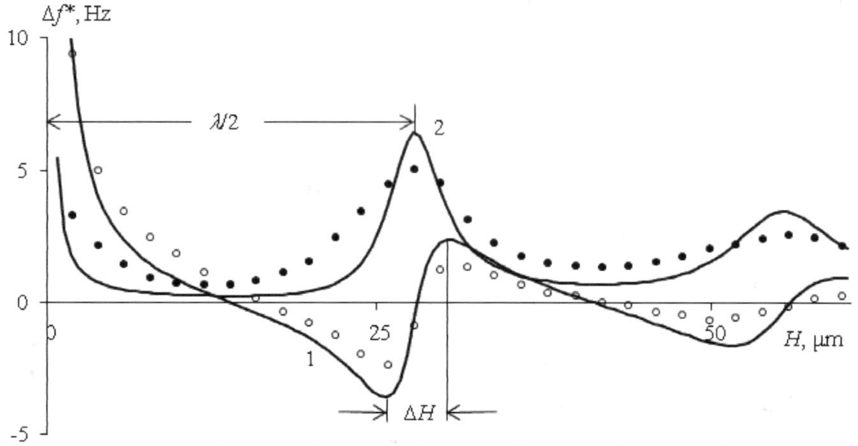

Figure 2. Theoretical (solid lines) and experimental (points) dependences frequency shifts on the thickness of the liquid layer for SiO_2/PES-2: 1 – real, 2 – imaginary frequency shifts.

4. Conclusion

Thus, the propagation of a low-frequency (10^5 Hz) shear wave in a colloidal suspension of SiO_2 / PES-2 was studied for the first time using an ultrasonic interferometer for shear waves. The shear wave length and the tangent of the mechanical losses angle were used from extreme values of the real and imaginary frequency shifts to calculate the value of the shear modulus of the investigated nanosuspension. The agreement of the experimental results obtained by the two ways of the acoustic resonance method confirms that the shear elasticity of the investigated nanosuspension is a property of the nanosuspension in the bulk.

Acknowledgment

The work was carried out within the framework of the state task of the IPMS SB RAS and partially supported by the RFBR grant, project No. 18-48-030020 r_a.

References

[1] Rudyak V Ya 2013 *Advances in Nanoparticles* **2** 266–79.
[2] Hey Y, Chen H, Ding Y, Cang D and Lu H 2007 *Int. J. Heat Mass Transfer* **50** 2272–81
[3] Chevalier J, Tillement O and Ayela F 2007 *Appl. Phys. Lett.* **91** 233 103
[4] Makarova D N, Dembelova T S and Badmaev B B 2020 *Acoust. Phys.* **66** 613–5
[5] Dembelova T S, Balzhinov S A, Makarova D N, Vershinina Ye D, Bazarova S B and Badmaev B B 2020 *IOP Conf. Ser.: Mater. Sci. Eng.* **1000** 012005
[6] Dembelova T S, Badmaev B B and Makarova D N 2019 *J. Phys.: Conf. Ser.* **1281** 012009
[7] Badmaev B B, Dembelova T S, Makarova D N and Vershinina E D 2018 *BSU bulletin. Mathematics, Informatics* **4** 37–47
[8] Badmaev B B, Budaev O R and Dembelova T S 1999 *Acoust. Phys.* **45** 610–4
[9] Badmaev B B, Dembelova T S, Makarova D N and Gulgenov Ch Zh 2020 *Russ. Phys. J.* **62** 1708–15
[10] Sobolevsky M V, Muzovskaya O A and Popeleva G S 1975 Properties and Applications of Organosilicon Products ed M V Sobolevsky (Moscow: Chemistry) p 296

IOP Conf. Series: Materials Science and Engineering 1198 (2021) 012002 doi:10.1088/1757-899X/1198/1/012002

Artificial and natural electromagnetic structures with strongly inductive surface impedance

Yu B Bashkuev, M G Dembelov, D G Buyanova, L Kh Angarkhaeva and I B Naguslaeva

Institute of Physical Materials Science of the Siberian Branch of the Russian Academy of Sciences, 6 Sakhyanovoy Street, Ulan-Ude 670047, Russia

buddich@mail.ru

Abstract. Artificial and natural electromagnetic structures with strongly inductive surface impedance are considered. The object of the research is layered structures of the "dielectric-conductor" type. The existence of impedance media with the maximum possible phase of a strongly inductive impedance in the range from a few hertz to tens of gigahertz has been established. The results of numerical modelling of the propagation of decimeter radio waves over a plane strongly inductive surface are presented, these numerical results are necessary for calculating the attenuation function W and the field level E of microwave electromagnetic waves.

1. Introduction

Due to the problem of surface electromagnetic waves (SEW), the study of the propagation of microwave radio waves over layered structures "dielectric – conductor" is of great importance in radiophysics. SEWs were first noted in the work of A. Sommerfeld [1]. From the formal point of view, SEWs are described by ordinary wave equations with standard boundary conditions, but they are their special solutions. At present, research on SEWs is developing quite dynamically in optic and microwave range [2,3]. In recent years, there has been an increased interest in SEW excitation at the interface between air and media with high conductivity (semiconductors and metals). These SEWs are also called surface plasmon polaritons (SPP). It should be noted that there is an electrodynamic commonality between SPP and the propagation of an electromagnetic wave over a strongly inductive impedance medium. Theoretical studies of SEW abroad were carried out by J. Wait [4]; in Russia, studies of the propagation of radio waves over layered impedance media were carried out at the Department of Radiophysics of Leningrad State University under the guidance of prof. G.I. Makarov [5].

S.A. Shchelkunov was one of the first scientists to propose an impedance approach to solving problems of electrodynamics and the theory of electrical circuits. [6]. The concept of impedance can refer to an electrical circuit, a medium, an interface, or simply a conductor. The work [7] presents an analytical review of the reference data on the use of the impedance approach in solving boundary value problems of electrodynamics for a 75-year period after M.A. Leontovich determined impedance boundary conditions for an electromagnetic field on the surface of a conducting body. In the calculations, the underlying medium must meet the impedance boundary conditions of Leontovich, that is, the condition $|\delta|^2 << 1$ must be carried out, where δ is the normalized surface impedance of the radio path. Many problems in radio physics and radio engineering (propagation and excitation of electromagnetic waves, shielding and grounding, absorbers of electromagnetic energy, etc.) are associated with calculations of the electrodynamic characteristics of layered media. Problems of this type include: 1) calculations of the surface (input) impedance of multilayer media, taking into account the absorption and dispersion properties of materials; 2) creation of electromagnetic coatings with specified characteristics of the transmission and reflection coefficients in a wide range of frequencies and angles of incidence; 3) finding the parameters of a layered medium from the measured frequency dependence of the surface impedance.

The purpose of the article is to search artificial and natural electromagnetic structures with the maximum possible phase of a strongly inductive impedance in the class of layered models of the medium. Layered models of some natural and artificial media with strongly inductive impedance are considered.

2. Method for calculating the surface impedance of a horizontally layered medium

To describe the electromagnetic properties of a layered medium, we have adopted a model of the structure of the medium, based on the idea of the presence of electrodynamically homogeneous areas. The vertical electrical section is assumed to be horizontally layered within each area. Each layer is characterized by conductivity $\sigma_j = 1/\rho_j$ (specific electrical resistance ρ_j), relative dielectric constant ε_j and thickness h_j. The horizontally layered model of the medium is the basic fundamental model in radiophysics. The advantage of this model lies in its simplicity, on the one hand, and in the possibility of an adequate description of real radiophysical processes, on the other hand.

The software has been developed. It includes programs: 1) "Impedance" for calculating the surface impedance of layered media and statistical calculations of impedance in a given frequency range; 2) "Inverse problem" for the interpretation of radio impedance soundings [8]. The calculation technique is based on the concept of the normalized surface impedance of a plane wave on the surface of a horizontally layered structure. With the help of impedance it is possible to take into account the influence of the underlying medium on the propagation of radio waves [4,5,8]: $\delta = E_\tau / (H_\tau \cdot Z_0)$, where E_τ, H_τ – horizontal mutually perpendicular components of the electric and magnetic fields at the air-ground interface; $Z_0 = \sqrt{\mu_0/\varepsilon_0} = 377$ Ohm is the characteristic impedance of vacuum.

The expression for the normalized surface impedance of a multilayer plane-parallel structure was used for calculations [8]:

$$\delta = \delta_1 \frac{1 - R_{12}^{(n)} \exp(-i2k_{1z}h_1)}{1 + R_{12}^{(n)} \exp(-i2k_{1z}h_1)},$$

where $R_{j(j+1)}^{(n)} = \dfrac{\delta_j \alpha_{j+1}^{+} - \delta_{j+1}\alpha_{j+1}^{-}}{\delta_j \alpha_{j+1}^{+} + \delta_{j+1}\alpha_{j+1}^{-}}$ is the reflection coefficient at the interface between the j-th and $(j+1)$-th layers; $\alpha_j^{\pm} = 1 \pm R_{j(j+1)}^{(n)} \exp(-i2k_{jz}h_j)$; $k_{jz} = k_0\sqrt{\varepsilon_{jk} - \sin^2\Theta}$; $k_0 = 2\pi/\lambda$; $\delta_j = \sqrt{\varepsilon_{jk} - \sin^2\Theta}/\varepsilon_{jk}$; at $j = n$ $R_{n(n+1)}^{(n)} = 0$, $\alpha_n^{\pm} = 1$; $\varepsilon_{jk} = \varepsilon_j + i60\lambda\sigma_j$ is the relative complex dielectric constant; h_j is the thickness of the j-th layer; Θ is the angle of incidence of a plane vertically polarized wave on the air-ground interface. It is assumed that the monochromatic electromagnetic wave has a time dependence $\exp(-i\omega t)$. The above relations make it possible to calculate the frequency dependences of the surface impedance of multilayer media.

3. Surface impedance of some types of layered media

Surface impedance depends on the structure of the underlying medium [4,5,7,8]. The following impedance structures are distinguished [5]:

- strongly inductive ($\mathrm{Im}\,\delta < 0, |\mathrm{Im}\,\delta| > \mathrm{Re}\,\delta$);
- low inductive ($\mathrm{Im}\,\delta < 0, |\mathrm{Im}\,\delta| < \mathrm{Re}\,\delta$);
- low capacitive ($\mathrm{Im}\,\delta > 0, \mathrm{Im}\,\delta < \mathrm{Re}\,\delta$);
- strongly capacitive ($\mathrm{Im}\,\delta > 0, \mathrm{Im}\,\delta > \mathrm{Re}\,\delta$).

It is known [5] that two-layer media of the "insulator on a conductor" and "conductor on an insulator" type with highly contrasting values ρ_j, ε_j allow changing the impedance from strongly inductive to strongly capacitive values ($-90° < \varphi_\delta < +90°$). According to the theory of radio wave propagation [4,5], the field over a strongly inductive path can exceed the field over an infinitely conducting plane at some distances. The interference between spatial and surface waves also causes field oscillations. A low strongly inductive subdomain ($-45° > \varphi_\delta > -72°40'$) and a strongly strongly inductive subdomain ($-90° < \varphi_\delta < -72°40'$) are distinguished, in which there are additionally "critical" impedance phases $\varphi_{2cr} = -80°$ and $\varphi_{3cr} = -83°30'$. All these cases are realized only for $|\varepsilon'_{1c}| < |\varepsilon'_{2c}|$.

In the class of natural layered media, an example of a strongly inductive structure is a two-layer "ice-sea" medium, which occupies significant areas of the World Ocean (figure 1). Ice-covered water areas meet the impedance boundary conditions in the VLF-LF-MF range, since the condition $|\delta|^2 << 1$ is carried out.

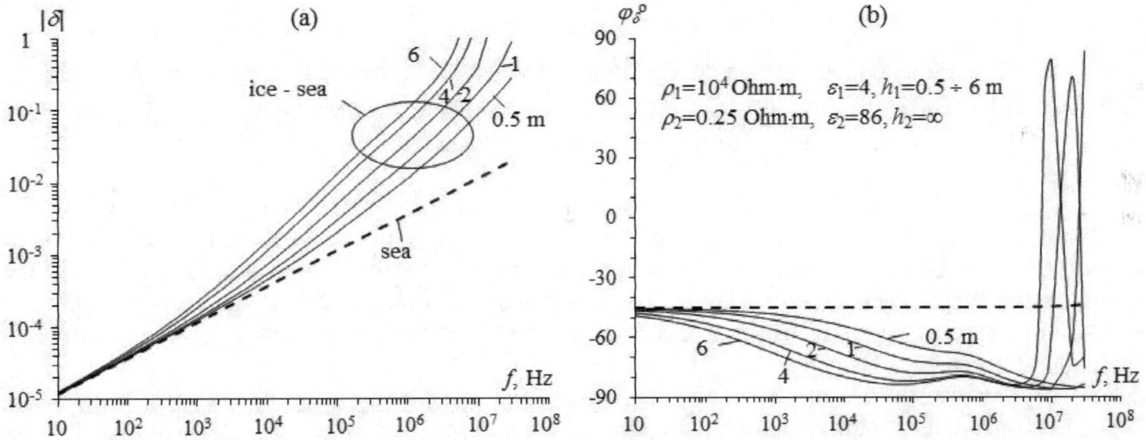

Figure 1. Frequency dependences of the modulus $|\delta|$ (a) and phase φ_δ (b) of the surface impedance of a two-layer medium "ice-sea". The numbers on the graphs are the thickness of the ice layer.

The geoelectric sections (GES) obtained on the sea sandy beaches of Asia washed by the waters of the Indian Ocean are presented in [9]. Sedimentary rock is a layered medium "dry sand - watered sea sand", most often consisting of 3-4 layers. The generalized geoelectric section represents the "dielectric on a conductor" structure, which is very similar in its electrical and geometrical properties to the "ice-sea" medium we have previously considered. Calculations of the surface impedance frequency dependences of the layered medium "dry sand - watered sea sand" in the range of 10 kHz - 30 MHz confirmed the very close electrodynamic similarity of such structures. The main difference between such structures is the stongly inductive impedance in the range from 10 kHz to 5-10 MHz. As an example, two three-layer GESs and their frequency dependences of the surface impedance of the medium “dry sand - watered sea sand” in the range of 10 kHz - 30 MHz are presented in figure 2. The “dry sand - watered sea sand” medium is analogous to the “ice-sea” structure in the VLF-HF range. (figure 2).

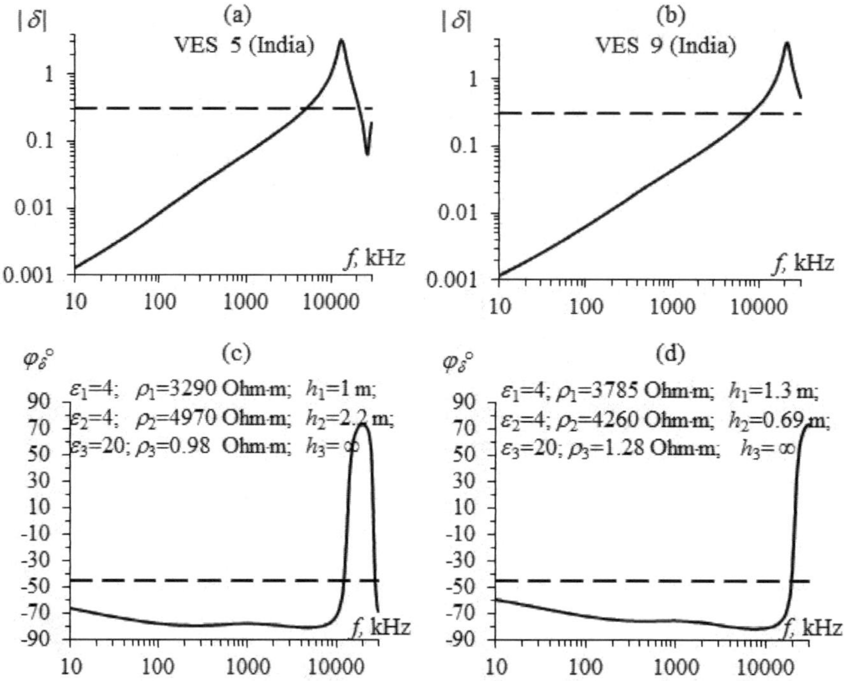

Figure 2. Frequency dependences of the modulus $|\delta|$ (a, b) and phase φ_δ (c, d) of the surface impedance for two GES of the “dry sand - watered sea sand” type: VES 5 (a, c) and VES 9 (b, d). The range of frequencies is 10 kHz - 30 MHz.

As a result of the search carried out in natural media, the existence of layered media with the maximum possible phase of strongly inductive impedance in the range from a few kilohertz to tens of megahertz was established. Let's note the wide distribution of beaches on all oceans and seas of the Earth. Consequently, such structures are typical for large areas of the Earth's surface and must be taken into account when calculating coastal refraction of the “land-sea” type in radio systems (direction finding and energy).

Let us consider the results of modelling the surface impedance of a thin three-layer medium “dielectric on a metal magnetic base” in a wide frequency range from 1 Hz to 10 GHz, corresponding to the real structure “paint layer on a steel sheet” (figure 3). Magnetic permeability of the substrate will be changed from 1 to 100 and up to 10000. Calculations show that a two-layer dielectric film “priming-paint” with a thickness of 0.5 mm shifts the impedance phase into the strongly inductive region already at frequencies from 100 Hz and above to 10 GHz. The impedance modulus satisfies the impedance boundary conditions over the entire frequency range. The impedance phase shows that in the class of layered natural and artificial media there are media with the maximum possible impedance phase values of strongly inductive impedance close to –90°.

The obtained values of the surface impedance indicate that the conditions for the propagation of electromagnetic waves over such structures in a wide range of radio waves will have features due to the appearance of a surface electromagnetic wave (SEW). Based on the experimentally proven fact of the existence of a SEW [10] and the principle of electromagnetic similarity, the obtained calculations of the surface impedance make it possible to theoretically substantiate the existence of an SEW over the “dielectric on a conductor” structure in the decimeter range of radio waves.

IOP Conf. Series: Materials Science and Engineering 1198 (2021) 012002 doi:10.1088/1757-899X/1198/1/012002

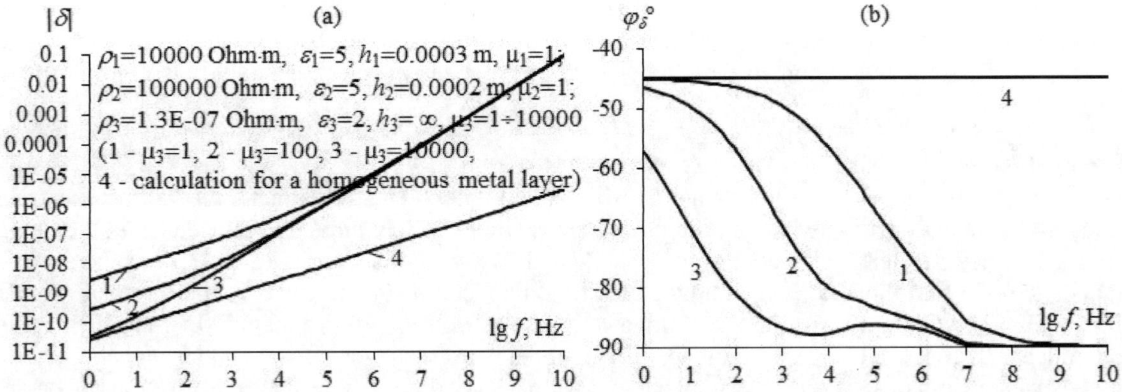

Figure 3. Frequency dependences of the modulus $|\delta|$ (a) and phase φ_δ (b) of the surface impedance of a three-layer medium "dielectric on a metal magnetic base".

Numerical modelling of the propagation conditions of radio waves over a model hypothetical homogeneous radio path for a GSM 1800 MHz cellular communication frequency showed the appearance of a SEW over the "dielectric – conductor" structure. Figure 4 shows the calculated values of the modulus of the field attenuation function $|W|$ and the field level E above a flat strongly inductive surface "dielectric on a metal magnetic base" for radiation power of 1 W.

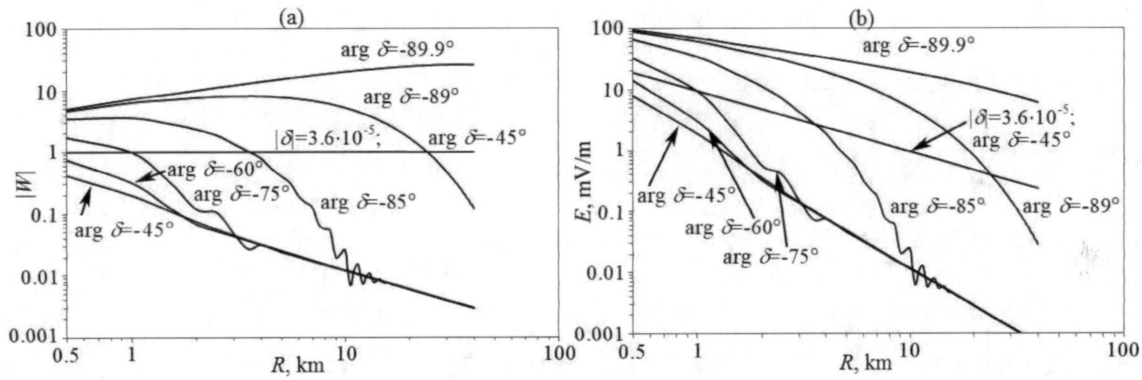

Figure 4. Calculated values of the modulus of the field attenuation function $|W|$ (a) and the field level E (b) above a plane strongly inductive surface "dielectric on a metal magnetic base" at a frequency of 1800 MHz. The radiation power is 1 W.

It can be seen from the graphs that in layered media with strongly inductive impedance, it is possible to achieve a large, purely electrodynamic in nature, amplification of the electromagnetic field. A strong dependence of the field amplification and the SEW propagation distance on the impedance phase has been established. Modelling has shown that the propagation of SEW in the decimeter range is better at closer distances from the emitter. That is, the presence of a two-layer dielectric film "primer-paint" with a thickness of 0.5 mm increases the field near the surface at short distances and decreases the field near the surface at long distances relative to the conductor. At the considered frequency, interference is observed between the spatial and surface waves, which causes oscillations of the field level. In practical terms, the effect of the appearance of SEW and interference can take place during the propagation of radio waves not only in cellular systems (GSM and CDMA) in the 450, 900 and 1800 MHz ranges, WiFi (2300 MHz), but also in the GPS radio navigation system L1 = 1575.42 MHz and L2 = 1227.6 MHz and GLONASS over real structures, for example, over traffic jams (fractal radio path). It is necessary to carry out additional experiments and propose new technical solutions based on the effect of the appearance of SEW for radio systems of the UHF-microwave ranges. Methods for measuring surface impedance in the microwave range should also be improved [11].

4. Conclusion

As a result of the search carried out among artificial and natural electromagnetic media, the existence of layered structures with the maximum possible phase of a strongly inductive impedance in the range from a few hertz to tens of gigahertz was established. Based on the results of modelling the surface impedance of a thin three-layer medium "dielectric on a metal magnetic base" in the frequency range from 1 Hz to 10 GHz, it was found that a two-layer dielectric film "primer-paint" 0.5 mm thick shifts the impedance phase to a strongly inductive region already at frequencies from 100 Hz and above. The impedance modulus carries out the impedance boundary conditions over the entire frequency range. The phase impedance is close to –90°. It is shown that there is a SEW above strongly inductive structures. This unique electrodynamic process played and plays an important, and sometimes a decisive role in natural physical and physico-chemical processes on the surface of layered media in a wide range of electromagnetic waves from millihertz to terahertz. The accumulated data on SEWs make it possible to obtain new fundamental knowledge, the results of which can be used on real radio links. The introduction of new results can be carried out both for new microwave radio systems and for complex technologies, for example, in the field of metamaterials.

Acknowledgments
The article was prepared according to the state budget project 0270-2021-0004 "Development of the fundamental principles of the propagation of ELF-VLF-LF-MF-SF and VHF radio waves in inhomogeneous impedance channels" at the Institute of Physical Materials Science of the Siberian Branch of the Russian Academy of Sciences.

References

[1] Sommerfeld A 1909 *J. Ann der Physik* **10(4)** 665-736
[2] Knyazev B A and Kuzmin A V 2007 *Vestnik NSU. Series: Physics*. **2(1)** 105-22 (in Russian)
[3] Libenson M N 1996 *Soros Educ. J.* **10** 92-8 (in Russian)
[4] Wait J R 1962 *Electromagnetic Waves in Sratified Media* (New York-Paris: Pergamon Press)
[5] Makarov G I, Novikov V V and Rybachek S T 1991 Propagation of Electromagnetic Waves over the Earth's Surface (Moscow: Nauka) (in Russian)
[6] Schelkunoff S A 1938 *Bell System Tech. J.* **XVII(1)** 17-48
[7] Berdnik S L, Penkin D Yu, Katrich V A, Penkin Yu M and Nesterenko M V 2014 *Radiophysics and Radioastronomy* **19(1)** 57-80 (in Russian)
[8] Angarkhaeva L Kh 2002 Impedance software package for solving problems of radio impedance sounding *Certificate of official registration of the computer program* 2002610893 (in Russian)
[9] Priju C P, Karthick P, Kumar A, Abdul Hameed E, Madhavan K and Narasimha Prasad N B 2014 *Int. Symp. on Integrated Water Resources Management (IWRM–2014) (Kozhikode, India)* pp 536–551
[10] Bashkuev Yu B, Khaptanov V B and Dembelov M G 2010 *Tech. Phys. Letters* **36(2)** 136-9
[11] Trunin M R 1998 *Phys.-Uspekhi* **41(9)** 843–63

IOP Conf. Series: Materials Science and Engineering 1198 (2021) 012003 doi:10.1088/1757-899X/1198/1/012003

Resonator method for measuring the polarizability of small components of composite materials

I O Dorofeev, G E Dunaevskiy and A Shansho

Department of Radio-Physics, National Research Tomsk State University

E-mail: idorofeev@mail.tsu.ru, proecs@mail.tsu.ru, ahmadsho1@gmail.com

Abstract. The creation of new composite materials with specified radio-frequency properties is often carried out using small-sized inclusions, while the polarizability parameters of these small bodies in electric and magnetic fields play a significant role. The paper considers the possibility of measuring the polarizability of small components of the composite in an open quasi-optical resonator. It is shown that the values of the electric and magnetic polarizability of small particles can be estimated from the shifts of the resonant frequencies that arise when a small scatterer is placed in the resonator.

1.Introduction

In the design of new composite materials intended for radio wave applications, such as ensuring electromagnetic compatibility, electromagnetic safety, and others, as a rule, particles of small sizes in comparison with the working wavelengths are used. In this case, the electrophysical parameters of these particles can have a decisive influence on the final characteristics of the created composite. In these cases, the important parameters of the particles included in the composite are the values of their electric and magnetic polarizabilities, the knowledge of which makes it possible to further simulate the scattering characteristics of the given composite in a given frequency range. Of interest is the possibility of direct measurements of the values of the polarizability of small particles used to create a composite at the frequencies of its further use. In this work, attention is focused on the resonator technique, in particular, on the use for these purposes of an open quasi-optical resonator (OR), which makes it possible to study elements of small dimensions in comparison with the wavelength and to carry out measurements separately in electric and magnetic fields. This technique is especially relevant in the terahertz frequency range, where there is practically no alternative to OR, and the demand for composites is quite high.

The proposed approach is based on multiple interactions of a quasi-optical beam with an object under study in an OR formed by concave spherical mirrors [1], which has a high intrinsic Q factor and free access to regions of high field concentration. Unlike the "waveguide" technique shown in Figure 1, a) where the probe beam interacts once with the small object under study, the use of ORs (Figure 1, b) ensures their multiple interaction, proportionally increasing the influence of the small scatterer on probing wave.

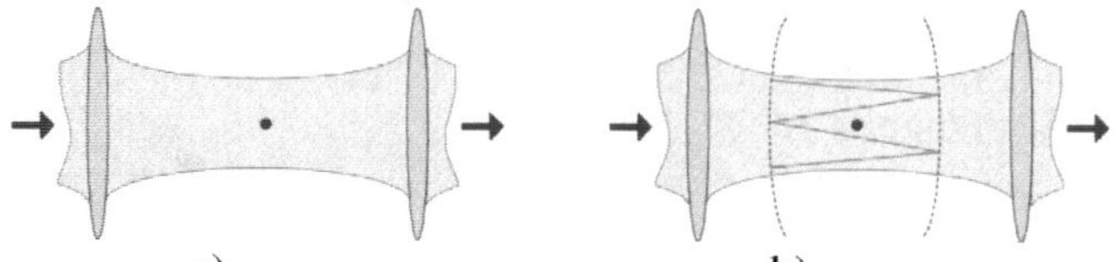

Figure 1. Single (a) and multiple (b) interaction of the probe beam with a small object.

A similar approach was used in [2, 3, 4], where the electrophysical parameters of aerogels from multi-walled carbon nanotubes, aerogel clusters, and sections of vitrified microwire with natural ferromagnetic resonance were measured. The main difficulty of these measurements was the problem of a sufficiently strong perturbation of the OR by objects and the need to search for restrictions for the correct application of the mathematical model of the method. The difference in this work is that the emphasis here is on measuring the polarizability of very small objects that disturb the OR, at the limit of the possibilities to observe changes in its spectral characteristics.

2. Description of the technique.

The placement of a small scattering device into the OR leads to shifts in the resonant frequencies of the resonator and changes in their Q-values [6,7]. If we assume that the small scattering device under study is located in the antinode of the electric field , then the electric dipole moment acquired by it will be described by the relation, where is the electric polarizability of this scattering device. Similarly, when placing it in the antinode of the magnetic field, we get where the magnetic polarizability is.

It is essential that both of these parameters of polarizability are determined by the shape, orientation of the sample in the field, the ratio of its size to the wavelength, the values of electric and magnetic permeability, which, in turn, are also frequency-dependent.

At the same time, it is possible to obtain the values of the electric and magnetic polarizability at a given frequency for a small body whose dimensions are much smaller than the wavelength experimentally, using the relations for the displacement of the resonant frequencies caused by the placement of the object under study in the field of an open resonator (here s is the oscillation index {mnq} OR):

$$\Delta f_s = -\frac{f_s}{N_s}(\varepsilon_0 \alpha_e E_s^2 - \mu_0 \alpha_m H_s^2), \tag{1}$$

where $N_s = \frac{\pi}{4} w_o^2 \varepsilon_0 L E_s^2$ is the norm of the s-th oscillation, is the radius of the "field spot" in the cross section of the location of the lens, L is the distance between the mirrors. When using a symmetric OR of a biconcave geometry with spherical mirrors with a radius of concavity ρ, the field spot in the center has a radius of [7]:

$$w_0 = \left(\frac{L\lambda}{2\pi}\right)^{1/2} \left[\frac{2\rho}{L} - 1\right]^{1/4}. \tag{2}$$

It is shown in [5,6] that the above relations, when applied to small metal spheres and to the duplex of small spheres, correspond satisfactorily to the experimental result.

3. Basic ratios

Let's use the above expressions to get the values of polarizability. From these relations, we obtain for the case of placing the diffuser in the antinode of the electric field, $(H_s = 0)$, [8]:

IOP Conf. Series: Materials Science and Engineering 1198 (2021) 012003 doi:10.1088/1757-899X/1198/1/012003

$$\alpha_e = \frac{\pi w_0^2 L}{2} \frac{\Delta f_{se}}{f_s}. \tag{3}$$

At the same time, in order to obtain a high resolution in the measurements of the values of the electric polarizability components, it is necessary to obtain as noticeable changes as possible in the measured parameters of the resonator (1) when a small diffuser is introduced into it. To clarify these conditions, we use the relations (2) and (3), we get

$$\frac{\Delta f_{se}}{f_s} = 4\lambda^{-1} L^{-2} \left(\frac{2\rho}{L} - 1 \right)^{-1/2} \alpha_e. \tag{4}$$

Acting similarly, we obtain for the magnetic polarizability measured by placing a small scatterer in the antinode of the magnetic field OR, ($E_s = 0$), [8]:

$$\alpha_m = \frac{\pi w_0^2 L}{2} \frac{\Delta f_{sm}}{f_s}. \tag{5}$$

And taking into account the expression (2), we get:

$$\frac{\Delta f_{sm}}{f_s} = 4\lambda^{-1} L^{-2} \left(\frac{2\rho}{L} - 1 \right)^{-1/2} \alpha_m. \tag{6}$$

Relations (4) and (6) describe the relationship between the measured values of the resonator parameters and the studied polarizability of a small body with different parameters of the geometry of the resonator used.

We introduce a dimensionless quantity:

$$A = \left(\frac{L}{2\rho} \right)^{-2} \left(\frac{2\rho}{L} - 1 \right)^{-1/2}. \tag{7}$$

Using this function, expressions (4) and (6) can be written as:

$$\frac{\Delta f_{se}}{f_s} = \frac{1}{\lambda \rho^2} A \alpha_e, \tag{8}$$

$$\frac{\Delta f_{sm}}{f_s} = \frac{1}{\lambda \rho^2} A \alpha_m. \tag{9}$$

In all these ratios, the value *A* shows the effect of the distance between the mirrors on the sensitivity of the technique. The radius of concavity of the mirrors is assumed to be fixed, which is quite consistent with the real situation.

Figure 2 shows the dependence of the value *A* on the ratio $L/2\rho$.

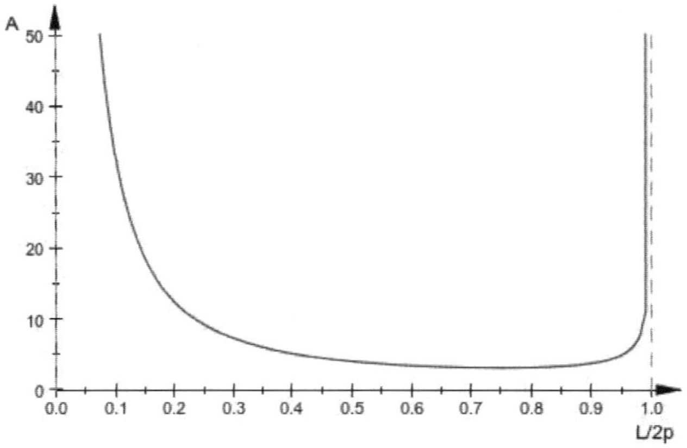

Figure 2. Dependence of the dimensionless parameter A on the distance between the resonator mirrors.

From the shown in Figure 2 dependences it follows that high values of parameter A can be obtained either when the mirrors approach at a distance less than 0.5 ρ , ($L/2\rho < 0.25$), or when the resonator configuration is close to concentric, ($L/2\rho \to 1$). The latter option is related to the fact that, according to expression (3), the value w_0 here approaches zero. However, in reality, such a strong reduction does not occur, since the diameter of the "field spot" in the focus cannot be less than the diffraction limit of 1.5 λ. In addition, in a concentric resonator, at $L/2\rho \to 1$, the size of the field spots on the mirrors increases sharply and, accordingly, the Q-factor decreases [7].

Excessive convergence of mirrors is also undesirable, since it is accompanied by a decrease in the quality factor of the resonator and the appearance of a large number of higher types of vibrations. It would seem that the increase in the value of A can be achieved not only by reducing L, but also by increasing the value of ρ, however, in the expressions (8), (9), the value ρ is present in the denominator, and so it is also impractical to increase it. As a result, we find that in the problem we are considering, it is inefficient to use mirrors with large concavity radii and located at large distances, although it is such OR that are most often used to obtain high Q-values. It follows from Figure 2 that the configuration used in many measurement problems, which is close to the confocal one ($L/2\rho \to 0.5$), is not optimal in this case.

It should be noted that the use of expressions (1) ÷ (6) to find the polarizability parameters of small bodies placed in the OP field, based on the results of measurements of the resonator parameters, is limited, on the one hand, by the size of the object under study (it must obviously be much smaller than the wavelength), and on the other – by the errors in the measurements of the displacement of the resonance curve and its half-width. The smaller the size of the small lens under study, the more accurately the given ratios will "work" (their good correspondence to the experiment is shown in [5,8]). But at the same time, the requirements for the accuracy of measurements of the resonator parameters without the object under study, and with it, will increase.

The requirements for the size of the resonator used can be illustrated by the example of the inclusion of a small dielectric sample with a shape close to spherical. For a small sphere of radius r_0 we get: $\alpha = r_0^3(\varepsilon - 1)/(\varepsilon + 2)$. Using the relation (9), we find:

$$\frac{\Delta f'_{se}}{f_s} = \frac{r_0^3}{\lambda \rho^2} A \frac{(\varepsilon - 1)}{(\varepsilon + 2)}. \tag{11}$$

Writing the radius of concavity of the mirror on the scale of the wavelengths used: $\rho = n\lambda$, for materials with high permittivity, ($\varepsilon >> 1$), we obtain for the minimum radius of the fixed particle $r_{\min}$ the ratio

$$\frac{r_{\min}}{\lambda} \approx \left(\frac{n^2}{A} \frac{\delta f_s}{f_s} \right)^{1/3}, \qquad (12)$$

where the value of δf is determined by the measurement error of the resonant frequency shift of the resonator. For example, with A value varying from 10 to 20, and n values ranging from 30 to 100, the ratio $\left(n^2 / A\right)^{1/3}$ will vary from 3.6 to 10 units, and the ratio of the minimum recorded radius of the particle to the wavelength will be almost completely determined by the possibility of fixing a small shift of the resonant frequency $\delta f / f$

Figure 3 shows the dependence of the minimum fixed radius of a small sphere (in fractions of the wavelength λ) on the parameter n (radius of concavity in wavelengths) for different distances between mirrors (dimensionless parameter A)and for different values of the value $\delta f / f$.

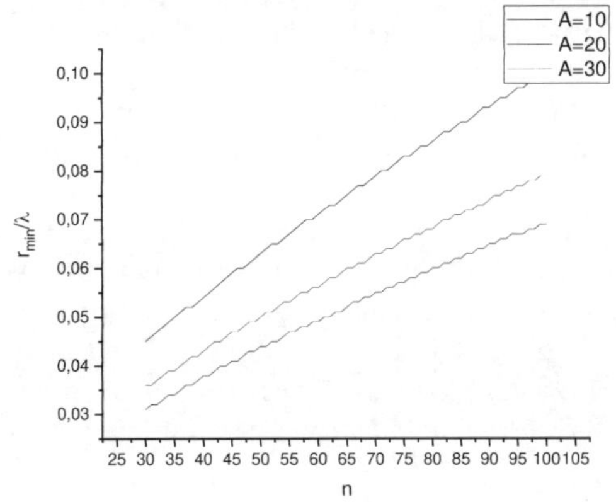

a) b)

Figure.3. The dependence of the minimum radius of the ball (in wavelengths) on the value of n at different values of the parameter A (a) and different frequency resolutions (b).

From the numerical estimates shown in Figure. 3, it is possible, with the optimal choice of the resonator size (values n of the order of 30-50, values A of the order of 20-30,) and the use of frequency synthesizers with a stability (step) of the order of $10^{-7} \div 10^{-8}$ and higher, to measure the parameters of small bodies with a radius of the order of several hundredth parts of the wavelength used.

4. Conclusion

Thus, using the quasi-optical resonator technique, it is possible to measure the values of the polarizability of small particles included in the composition of new composite materials. Measurements can be made at the frequencies of their further use. As for the possibility of reducing the size of the controlled particles, they will largely depend on the mechanical and temperature stability of the OR used and the stability of the vibration sources used. In the gigahertz range, digital circuit analyzers can be used for this purpose [9], and in terahertz it is advisable to use BWO generators, but they require additional stabilization by external frequency synthesizers [10]. The use of the OR allows to measure the polarizability separately in the electric and magnetic fields, which is important when evaluating the scattering properties of small elements of a composite material.

IOP Conf. Series: Materials Science and Engineering 1198 (2021) 012003 doi:10.1088/1757-899X/1198/1/012003

References

[1] Dicke R H 1958 Molecular amplification and generation systems and methods. U.S.Patent 2851652

[2] Kuznetsov V L, Suslyaev V I, Kazakova M A, Moseenkov S I, Smirnova T E and Krasnikov D V 2015 Investigation of electromagnetic properties of MWCNT aerogels produced via catalytic ethylene decomposition. *Phys. Status Solidi* B 52, No. 11, 2519–2523

[3] KrasnikovD V, Dorofeev I O, Smirnova T E, Suslyaev V I, Kazakova M A, Moseenkov S I and Kuznetsov V L 2017 Electromagnetic Interaction Between Spherical Aerogels of Multi-Walled Carbon Nanotubes *Phys. Status Solidi* B, 1700256.

[4] Dorofeev I O, Dunaevskii G E 2017 Resonance Characteristics for Microwire Pieces as Elements of Composite Materials *Russian Physics Journal*. 59, No. 12, April.

[5] Dunaevskiy G and Dorofeev I 2017 From Micro- to Nano-objects and from Giga- to Terahertz Frequency Range: Quasi-optical Resonant Diagnostics of Objects *Semiconductor Nanotechnology. Advances in Information and Energy Processing and Storage* Springer 169–198

[6] Tsipenyuk Yu M 1965 Test body in an open resonator *High-power electronics* 4 173-177

[7] Vainshtein L A 1966 *Open resonators and open waveguides*. Moscow Sovetskoe radio 474

[8] Dorofeev I O, Dunaevskiy G E, Smirnova T E, Shpilnoy VYu 2014 The test body method for the study of small objects in an open microwave resonator. *Izvestia Vuzov Physica* 57N9/2 35-40.

[9] Yang B B, Katz SL, Willis K J, Weber M J,.Knezevic I, Hagness S C and Booske J H 2012 A High-*Q* Terahertz Resonator for the Measurement of Electronic Properties of Conductors and Low-Loss Dielectrics. *IEEE Trans. On Terahertz Science And Technology* 2 N 4 July 449 – 459.

[10] Gao Y, Jones C R and Dutta J. M. 2009 Phase Locked BWO System for Open Resonator Measurement in D-Band *IEEE Microwave And Wireless Components Letters* 19 N 9 599 – 601.

IOP Conf. Series: Materials Science and Engineering 1198 (2021) 012004 doi:10.1088/1757-899X/1198/1/012004

Improving the shielding efficiency and weight reducing of the radio-absorbing coating by adding a matching layer

K V Dorozhkin[1], V I Suslyaev[1], O V Kazmina[2], K E Matyskin[3], E Yu Korovin[1]

[1] National Research Tomsk State University, 36 Lenina Ave., Tomsk 634050, Russian Federation

[2] National Research Tomsk Polytechnic University, 30 Lenina Ave., Tomsk 634034, Russian Federation

[3] Company «MK Electron», 43 Novosibirskaya St., Tomsk 634063, Russian Federation

E-mail: yasbtk@yandex.ru

Abstract. The spectra of complex permittivity of the composites consisting of rubber silicone compound and electroconductive fine-dispersed silver-coated copper powder were investigated. To enhance the effectiveness of the composite interaction with microwaves and reduction of its weight and dimensions characteristics two and three-layer constructions were produced. Matching layers were made of foam glass material with addition SiC and Fe_3O_4.

1. Introduction

Materials effectively shielding diverse objects are widely used in various sectors: radioelectronics, ecology, medicine, counter-terrorism systems and security [1-4]. The increased use of radioelectronic equipment in science and production, for improvement of living conditions [5] poses a number of problems. First, it pollutes the environment and contributes to increased incidence. In this regard, shielding devices are necessary for biological objects protection and separate blocks of radioelectronic equipment interference suppression. The screens are constantly being developed, but new challenges appear and the old methods become outdated. The use of well-established metal (steel, copper, aluminium etc.) screens is limited now due to miniaturization, when weight and size are of primary importance. One of the possible ways of weight reduction is production of composite materials made of conductive polymers [6] or usage of dielectric polymer matrices filling with conductive particles. If the filling exceeds the percolation threshold, the screen reflects electromagnetic energy, i. e. it becomes analogous to a metal screen but lighter by weight. If the filling is below the percolation threshold, non-reflecting screens where electromagnetic energy transforms to thermal can be produced. Polymer screens are flexible, elastic, inexpensive and producible. Nowadays polymer materials with fillings are used in 3D printer filaments production allowing to produce protective screens of the most complicated shapes [7,8]. In particular, 3D printing makes it possible to design metasurfaces, that allow matching the wave impedances of free space and a shielding device. [9,10]. Creation of new materials based on foaming [11,12] and carbonic aerogel synthesis [13, 14] leads to weight reduction.

2. Materials and methods

The composite consisting of rubber silicone compound IRP-1401 (technical specifications (TS) 38.103372-77) and electroconductive fine-dispersed silver-coated copper powder PMI-certified (TS 24.44.21-001-32946049-2018) was used as main shielding material. Compounding was performed on cold gap-mill with 1:1.15 friction coefficient. PMS-100 - certified silicone oil was used as an adjuvant improving metal powder incorporated into matrix. The amount of electroconductive powder in the compound is 85 wt %. By controlling the conductivity of the compound, the mixing quality was achieved. Further, electroconductive silicone compound was loaded into rubber vulcanizing press at clamping force 100 kg/cm. The time of vulcanization was 10 min, vulcanization temperature was 190°C. Plate was cooled at room temperature for 24 hours after extraction from the pressform. Finally, the composite (TSC) of 4,06 g/cm^3 density was obtained.

A foam glass material with addition of silicon carbide and ferrous oxide [15] was used as a matching layer having lowest real and imaginary component of permittivity, which provides reduction of characteristic impedance jump. Porous foam glass was produced by low-temperature method described in [15]. Sodium liquid glass of 2.5 acidity index was used as a matrix, where cullet powder of 6000 cm^2/g specific surface was incorporated. Hydrophilic aluminum powder was used as a gasifier. To improve bond performance SiC was added in liquid glass composite in the amount of 20 and 40 wt % instead of cullet powder. Silicon carbide was chosen due to large width of its forbidden gap (3.26 eV), permittivity (6.5-7.5 rel. un.) and covalent type of chemical bond, which influences on material strength properties.

During foam glass material synthesis, it was observed that the increase of SiC content the value of expansion coefficient increases. The reaction of additional gasification in the system results in porosity increase and composite density decrease (table 1). This effect is corroborated by scanning electron microscopy (SEM) pictures (figure 1).

Table 1. Physical-mechanical properties of the composites.

Composite	Amount of SiC in the composite, mass percent	Physical-mechanical properties	
		density, kg/m3	compression strength, MPa
PSSiC-20	20	388	1.43
PSSiC-40	40	354	0.50

When the cullet powder was used as a filler between liquid glass and glass particles as similar components, the surface interaction was observed. When initial materials were replaced by powders, greatly differed by its properties (density, strength, wettability), the cohesion between matrix and filler was decreased. If the concentration is rather low, silicon carbide particles strengthen the composite. Silicon carbide content above 40 wt % leads to a decrease in the strength of the final sample due to the formation of an uneven porous structure and an increase in the average pore size (figure 1 b). According to obtained experimental data the criteria for formation of even, finely porous structure in liquid glass composition were determined: the ratio of the mass of the liquid phase to the solid filler should be 0.35; dynamical viscosity of the composition must be at 1.5 Pa×s rate.

To increase the electromagnetic effectiveness of foam glass material, GaAs semiconductor material was incorporated in the structure of the filler. In doing so, the problem of recycling in order to decrease anthropogenic burden on the environment was solved. To estimate the possibility of this use of the technology 40 wt %. SiC was replaced by the waste of semiconductor industry (marking of the sample – PSSiC-GaAs). In attempt to use magnetic properties of the composites, ferrous oxide (marking – PSSiC-Fe_3O_4) was incorporated instead of silicon carbide in the same proportion as semiconductor material.

IOP Conf. Series: Materials Science and Engineering 1198 (2021) 012004 doi:10.1088/1757-899X/1198/1/012004

Complex permittivity and electromagnetic response components (reflection, transmission and absorption factors) of the composites were measured by STD-21 terahertz spectrometer [16]. Calculation of complex permittivity spectra was made for normal incidence of flat electromagnetic wave on flat material surface.

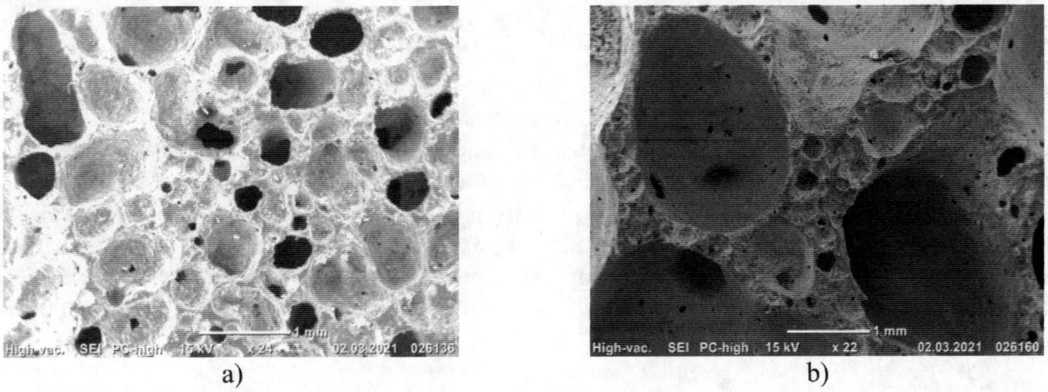

Figure 1. SEM pictures of composites with 20 wt % of SiC (a) and 40 wt % of SiC (b).

3. Results and discussion

Composite complex permittivity spectra of silver-coated copper powder in silicone matrix have weak decreasing dependence on frequency increase, while imaginary component have no dispersion (figure 2).

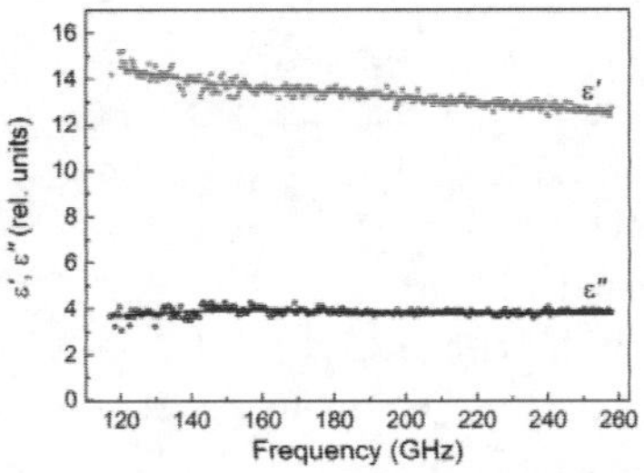

Figure 2. Complex permittivity spectrum of silicon composite TSC.
ε' is real part of permittivity, ε'' is imaginary part of the permittivity.

Foam glass material without additive (PS) has $\varepsilon^* = 2.2 - i \cdot 0.42$ without dispersion in 120-250 GHz bandwidth. Frequency dependence of permittivity complex values of foam glass composites with additives are shown on figure 3.

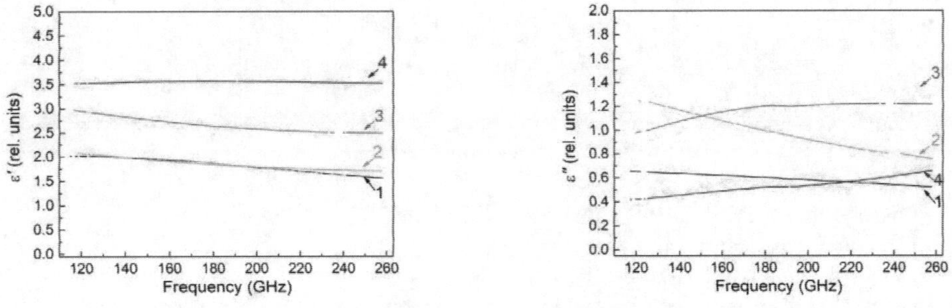

Figure 3. Complex values spectra of permittivity complex values of foam glass composites:
1 – PSSiC-20, 2 – PSSiC-40, 3 – PSSiC-GaAs, 4 – PSSiC-Fe_3O_4.

IOP Conf. Series: Materials Science and Engineering 1198 (2021) 012004 doi:10.1088/1757-899X/1198/1/012004

The change of silicon carbide concentration in the investigated bandwidth influenced on imaginary component and had no significant influence on real component. The addition of gallium arsenide to the foam glass led to an increase in both components of the permittivity. In this case, the imaginary part of the permittivity increases in the frequency range from 120 to 180 GHz with subsequent stabilization of the value in the region of 1.2 relative units in the frequency range from 180 to 250 GHz. The addition of ferrous oxide also raised the value of permittivity. All the examined materials have much weaker interaction with electromagnetic field than TSC composite and have closer values to the parameters of free space.

The frequency dependences of the electromagnetic response (transmission, reflection and absorption), obtained by solving the direct electrodynamic task, are shown in figure 4–6. Figure 4 demonstrates the characteristics of two-layer construction, where remote layer is 1 mm thick made of silicon composite with copper powder. Near to the front of incident wave layer was replaced by foam glass samples where silicon carbide concentration was 20 and 40 wt %, then by silicon carbide, gallium arsenide and ferrous oxide combinations. The thickness of matching layers was 2 mm. Electromagnetic characteristics of one-layer TSC composite are shown on the same figure for comparison.

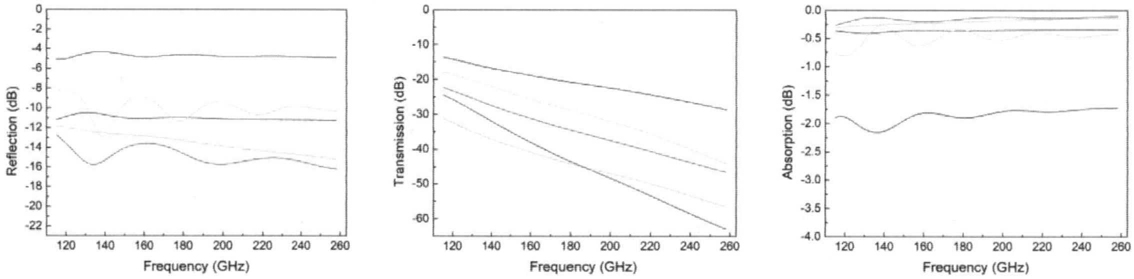

TSC (1mm)
TSC (1mm) + PSSiC-20 (2mm)
TSC (1mm) + PSSiC-40 (2mm)
TSC (1mm) + PSSiC-GaAs (2mm)
TSC (1mm) + PSSiC-Fe_2O_4 (2mm)

Figure 4. Two-layer coating electromagnetic response components.

As it is shown, a construction reflectance decreased significantly. Transmission also decreased due to the growth of construction absorbance.

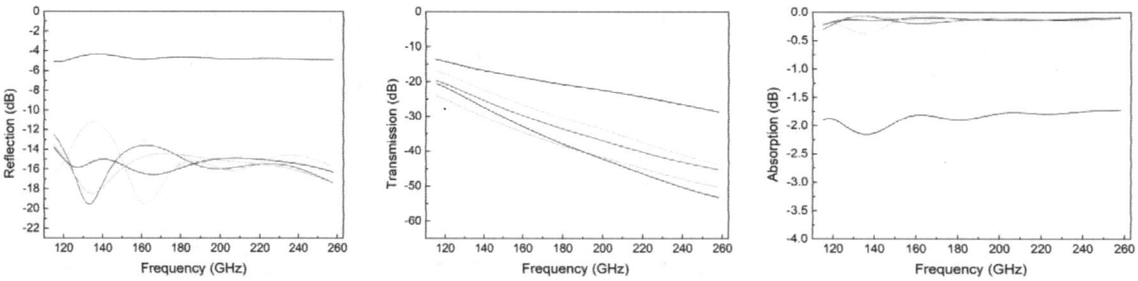

TSC (1mm)
TSC (1mm) + PSSiC-20 (1mm) + PS 1.1 (1mm)
TSC (1mm) + PSSiC-40 (1mm) + PS 1.1 (1mm)
TSC (1mm) + PSSiC-GaAs (1mm) + PS 1.1 (1mm)
TSC (1mm) + PSSiC-Fe_2O_4 (1mm) + PS 1.1 (1mm)

Figure 5. Three-layer coating electromagnetic response components.

The replacement of half-thickness of matching layers in the construction by foam glass material without additives 1 mm thick didn't change the whole thickness but significantly decreased reflectance ratio due to contrast decay between medium impedance bringing electromagnetic energy and three-layer structure.

Figure 6 shows electromagnetic response spectra of the construction, where remote layer thickness was reduced by half and the thickness and control positions of matching layers remain the same as on Figure 5.

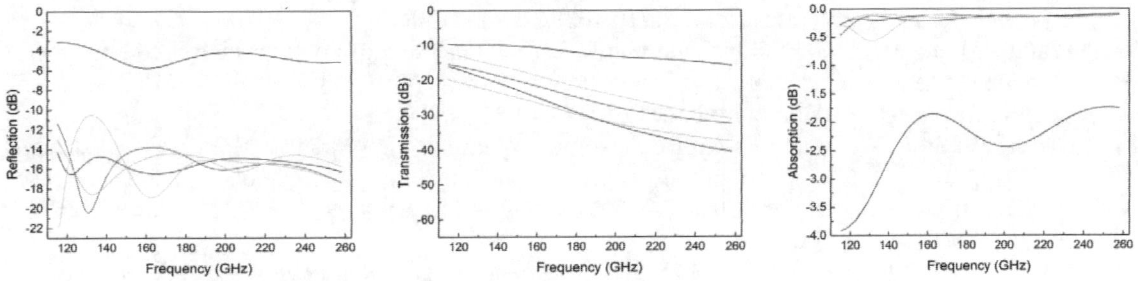

TSC (0.5mm)
TSC (0.5mm) + PSSiC-20 (1mm) + PS 1.1 (1mm)
TSC (0.5mm) + PSSiC-40 (1mm) + PS 1.1 (1mm)
TSC (0.5mm) + PSSiC-GaAs (1mm) + PS 1.1 (1mm)
TSC (0.5mm) + PSSiC-Fe_2O_4 (1mm) + PS 1.1 (1mm)

Figure 6. Three-layer coating electromagnetic response components with reduced thickness of first layer.

As can be seen (black line in figure 5 and figure 6), the reflectance remained approximately the same as for 1mm thick sample of TSC, transmission increased and absorption decreased. The problem was solved by addition of matching layers. Although absorption decreased, other characteristics of electromagnetic response enabled to use the construction for effective shielding.

4. Summary

The research has shown that electrophysical characteristics of foam glass material are alterable depending on filler composition. The construction made of main absorber (silver-coated copper powder in silicon matrix) and foam glass matching layer enables to decrease reflectance from –5 dB to –16 dB. The usage of matching layers reduced the weight of the construction by half, as the remote layer with the density of 4,06 g/cm^3 has the main weight, foam glass material being 10 times lighter.

5. Acknowledgement

This research was supported by Ministry of Science and Higher Education of the Russian Federation, project No 0721-2020-0038.

References

[1] Shukla V 2019 Review of electromagnetic interference shielding materials fabricated by iron ingredients *Nanoscale Adv.* **1** 1640–1671 https://doi.org/10.1039/c9na00108e

[2] Zhang Liying, Bi Shuguang and Liu Ming 2018 Lightweight Electromagnetic Interference Shielding Materials and Their Mechanisms *IntechOpen.* 1–20 https://doi.org/10.5772/intechopen.82270

[3] Micheli D, Morles R B, Marchetti M, Moglie F and Primiani Mariani V 2014 Broadband electromagnetic characterization of carbon foam to metal contact *Carbon* **68** 149–158 https://doi.org/10.1016/j.carbon.2013.10.074

[4] Grimaldi C, Kecsenovity E, Majidian M, Kuznetsov V L, Magrez A and Forro L 2019 Electrical transport in onion-like carbon - PMMA nanocomposites *Applied Physics Letters* **114** 103102 https://doi.org/10.1063/1.5085936

[5] Wang C, Murugadoss V, Kong J, He Z, Mai X, Shao Q, Chen Y, Guo L, Liu C, Angaiah S and Guo Z 2018 Overview of carbon nanostructures and nanocomposites for electromagnetic wave shielding *Carbon* **140** 696–733 https://doi.org/10.1016/j.carbon.2018.09.006.

[6] Yang J, Liu Y, Liu S, Li L, Zhang C and Liu T 2017 Conducting polymer composites: Material synthesis and applications in electrochemical capacitive energy storage *Materials Chemistry Frontiers* **1** 251–268 https://doi.org/10.1039/c6qm00150e

[7] Wang X, Jiang M, Zhou Z, Gou J and Hui D 2017 3D printing of polymer matrix composites: A review and prospective *Composites Part B: Engineering* **110** 442–458 https://doi.org/10.1016/j.compositesb.2016.11.034

[8] Gnanasekaran K, Heijmans T, van Bennekom S, Woldhuis H, Wijnia S, deWith G, Friedrich H 3D printing of CNT- and graphene-based conductive polymer nanocomposites by fused deposition modeling *Applied Materials Today* **9** 21–28 https://doi.org/ 10.1016/j.apmt.2017.04.003

[9] Gorokhov G V, Bychanok D S, Kuzhir P P, Gorodetskiy D V, Kurenya A G, Sedelnikova O V, Bulusheva L G and Okotrub A V 2020 Creation of metasurface from vertically aligned carbon nanotubes as versatile platform for ultra-light THz components *Nanotechnology* **31** 255703 https://doi.org/10.1088/1361-6528/ab7efa

[10] Wu K, Coquet P, Wang Q J and Genevet P 2018 Modelling of free-form conformal metasurfaces *Nature Communications* **9** 3494 https://doi.org/10.1038/s41467-018-05579-6

[11] Letellier M, Macutkevic J, Paddubskaya A, Klochkov A, Kuzhir P, Banys J, Fierro V and Celzard A 2015 Microwave dielectric properties of tannin-based carbon foams *Ferroelectrics* **479** 119–126 https://doi.org/10.1080/00150193.2015.1012036

[12] Chasnyk V I 2012 High absorption of the microwave energy in a system with strongly elongated molybdenum grains in aluminum nitride matrix at frequencies of 9.5-10.5 GHz *Journal of Superhard Materials* **34** 71–73 https://doi.org/10.3103/S1063457612010091

[13] Wang Z, Wei R, Gu J, Liu H, Liu C, Luo C, Kong J, Shao Q, Wang N, Guo Z, and Liu X 2018 Ultralight, highly compressible and fire-retardant graphene aerogel with self-adjustable electromagnetic wave absorption *Carbon* **139** 1126–1135 https://doi.org/10.1016/j.carbon.2018.08.014

[14] Dorofeev I O, Suslyaev V I, Moseenkov S I, Kuznetsov V L and Dorozhkin K V 2019 Interaction of Multiwalled Carbon Nanotube Aerogels with Quasi-Optical Terahertz Beams *Physica Status Solidi Basic Res.* **256** 1900251 https://doi.org/10.1002/pssb.201900251

[15] Kaz'mina O V, Suslyaev V I, Dorozhkin K V, Kaimonov M R and Stebeneva V I 2018 Effect of a coating deposited on foam glass on the ability to absorb high-frequency electromagnetic radiation *Glass and Ceramics* **75** 230–233 https://doi.org/10.1007/s10717-018-0061-2

[16] Dunaevskii G E, Suslyaev V I, Zhuravlev V A, Badin AV and Dorozhkin K V 2016 Ferromagnetic resonance in hexagonal ferrite $Ba_3Co_2Fe_{24}O_{41}$ at the THz frequency range I*nt. Conf. Infrared, Millimeter, Terahertz Waves, IRMMW-THz* **2016** 7758771 https://doi.org/10.1109/IRMMW-THz.2016.7758771

Electromagnetic characteristics of materials based on ultrafine hexaferrite powders

YuN Gladkov [1], AS Kachalov [2], EYu Korovyn [3] and AA Pavlova[4]

[1] Graduate student, National Research Tomsk State University, Tomsk, Russia
[2] Student, National Research Tomsk State University, Tomsk, Russia
[3] Associate professor, National Research Tomsk State University, Tomsk, Russia
[1] Graduate student, National Research Tomsk State University, Tomsk, Russia

E-mail: sandy.surname@gmail.com, susl@mail.tsu.ru

Abstract. The electromagnetic characteristics of composite materials based on nanosized powders of W-type hexaferrites are considered in the article. It is shown that not only the composition, but also the mechanical treatment affects the electromagnetic parameters. This article presents the frequency dependence of the complex magnetic and dielectric permittivity of a system of W-type hexafferites. The studies were carried out on a universal wide-band measuring complex based on the Agilent PNA-X N4257A Vector Network Analyzer. The results are presented in the frequency range from 2 to 14 GHz.

1. Introduction

The problem of developing magnetic radio-absorbing materials remains relevant at present time due to the expansion of the areas of their application, in particular, the active introduction of microwave radiation into household appliances, navigation systems, communications and computing [1-3]. A new impetus in the development of radio-absorbing materials is caused by a surge of interest among scientists and development engineers in nanosized powders of hexaferrites, the static properties of which are fundamentally different from the bulk material [4]. The relevance of the problem is confirmed by numerous conferences, seminars and symposia held by different countries and the growing flow of scientific publications in various journals on this topic.

Materials operating in a wide frequency range are required for radio-absorbing coatings, and coatings made of these materials must have: low thickness, low weight, good radio-absorbing properties, and negligible reflection of electromagnetic radiation. The electromagnetic compatibility of individual units of high-frequency radio electronics is ensured by the use of radio-absorbing materials, the effectiveness of modern communication means is increased, and the protection of users from the harmful effects of the microwave range is formed. Coatings based on magnetodielectric materials, which can be made in the form of solid samples or serve as fillers for elastomers, work most effectively on a metal surface with high electrical conductivity, where the antinode of the magnetic field is located.

Ferrites of various compositions, powders of iron and nickel carbonyls are used as magnetic materials. Ferrites with a hexagonal crystal structure (hexaferrites) compare favorably with materials that have high dispersion frequencies determined by the resonance of domain walls and natural ferromagnetic resonance (NFMR). A significant increase in the imaginary component characterizing the loss of electromagnetic radiation in matter is observed in these regions, along with a rather sharp decrease in the value of the real component of the complex magnetic permeability $\mu^* = \mu' - i\,\mu''$.

The system of W-type hexaferrites was selected as a specific object of research. This system is interesting as a model material for studying a fundamental phenomenon in magnetically ordered structures - a spin-orientation phase transition [5].

Hexaferrite powders were obtained using the following technology. Massive pieces of ceramic were ground in a mortar (initial powder). After that, mechanical activation was carried out in a planetary mill at a factor of energy intensity of 15g, which corresponds to a soft processing mode (in this case, the ratio of grinding balls to a sample by weight is 5: 1). The static characteristics (specific and remanent magnetization, the magnitude of the anisotropy field) of the synthesized nanosized materials are given in Table 1. The measurements were carried out according to the methods described in [6].

The resulting nanosized powder was compacted into a calibrated thin-walled quartz tube with a knows volume to determine the complex values of the magnetic permittivity. The studies were carried out on a universal wide-band measuring complex based on the Agilent PNA-X N4257A Vector Network Analyzer. The resonators were used as a measuring cell. They are built from waveguide elements of standard measuring lines, which are manufactured with great care at the factory. The block diagram of the measuring setup is shown in Figure 1.

Table 1. Specific and residual magnetization, dependence of the magnitude of the anisotropy field on the time of mechanical activation

Processing time, s	Specific magnetization σ, $Gs*cm^3/g$	Residual magnetization, $(Br/Bs)\times10^2$	Anisotropy field Ha, Gs
0	73.00	26	1000
30	69.25	9.76	1300
60	63.82	12.9	1555

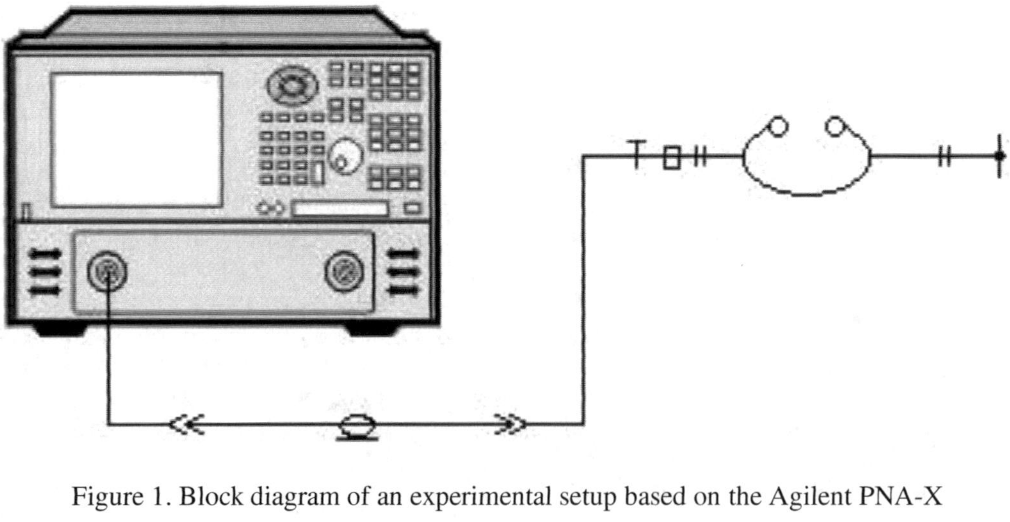

Figure 1. Block diagram of an experimental setup based on the Agilent PNA-X N4257A Vector Network Analyzer

IOP Conf. Series: Materials Science and Engineering 1198 (2021) 012005 doi:10.1088/1757-899X/1198/1/012005

2. Results and discussion

The main idea of the resonator method is to observe the resonance curve of the measuring cell into which the sample under study is introduced. The shift of the resonance frequency and the growth of the line width, which can be used to calculate, occurs when the sample is placed in the cavity of the cavity resonator. Multimode provides a relative broadband of the resonator method, and allows one to restrict oneself to a small number of measuring cells with sufficient resolution (on 7 resonators 120 points in the range from 2 to 14 GHz). The software described in [7, 8] was used to calculate the complex values of the permittivity and permeability.

The studies carried out have shown (Figure 2, 3) that the measured values significantly depend not only on the composition of the material, but also on the time of mechanical processing of nanosized powders.

It can be seen that in the region of 6 - 7 GHz, a pronounced EFMR is observed for materials subjected to machining. The EFMR frequency for this material increases by 16% with a change in time from 30 to 60 seconds with an increase in processing time. At the same time, a decrease in the value of the imaginary component of the magnetic permeability, which is associated with a decrease in the specific magnetization, is observed. A change in the EFMR frequency is associated with a change in the magnitude of the an anisotropy field. Thus, the investigated dynamic parameters (dielectric permittivity and magnetic permeability) are in good agreement with the measured static characteristics.

As expected, no features were found at the complex dielectric constant. The main mechanisms that form the electrical properties of ferrites are electronic $\omega_{ep}\approx 10^{16}$ and ionic polarizations $\omega_{ep}\approx 10^{13}$, the critical frequencies of which lie in the region of electronic transitions and optical branches of the phonon spectrum. Therefore, a weak frequency dependence of ε * is observed in the microwave range.

The calculation of the reflection coefficient of a normally incident electromagnetic wave on a plane-parallel surface was carried out according to the data obtained. The reflection coefficient itself was calculated using the well-known formula for the case when a radio-absorbing layer of nanosized hexaferrite powder is located on an ideally conducting surface.

$$R = 20\log\left|\frac{Z_{in} - Z_0}{Z_{in} + Z_0}\right| \tag{1}$$

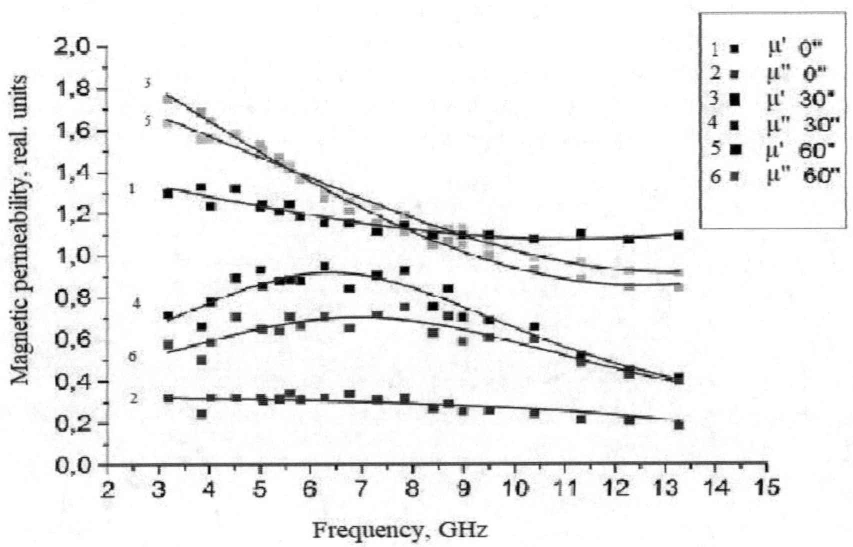

Figure 2. Frequency dependence of the complex permeability of nanosized powders of the hexaferrite system $Co_{0.7}Zn_{1.3}W$

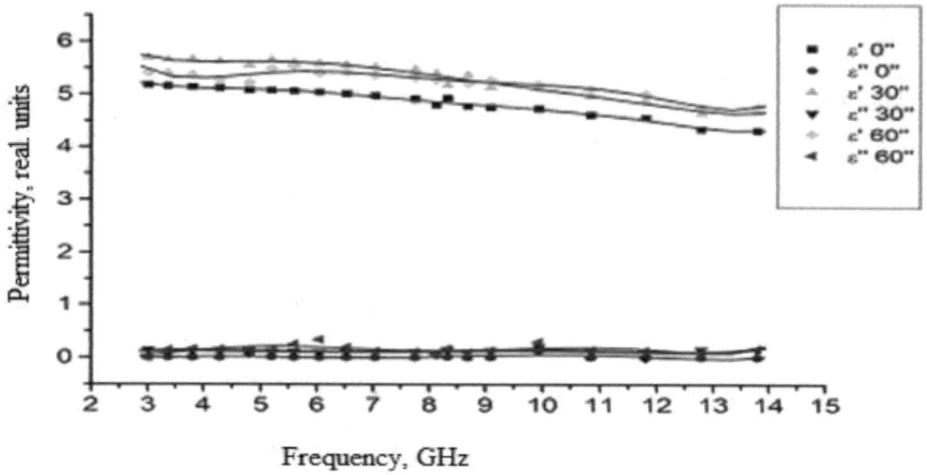

Figure 3. Frequency dependence of the complex permittivity of nanosized powders of the hexaferrite system $Co_{0.7}Zn_{1.3}W$

For the case when a radio-absorbing layer of nanosized hexaferrite powder is located on an ideally conducting surface. In formula 1 $Z_0 = \sqrt{\mu_o/\varepsilon_o}$ – is the wave number of free space, Z_{in} is the impedance at the interface between free space and absorbing material, determined by the formula

$$Z_{in} = \sqrt{\frac{\mu_0\mu}{\varepsilon_0\varepsilon}} \tanh\left(i2\pi f \sqrt{\mu_0\mu\varepsilon_0\varepsilon d}\right) \tag{2}$$

where f is the frequency of the electromagnetic field, d is the thickness of the material, μ, ε are the complex magnetic and dielectric permittivities of the material.

Figure 4 shows the frequency dependence of the reflection coefficient for nanosized powders with different activation times. It can be seen that the most broadband (at a level of 10 dB) is a sample with a 30 s processing time, the operating range of which lies in the 5.5 ÷ 10.5 GHz band.

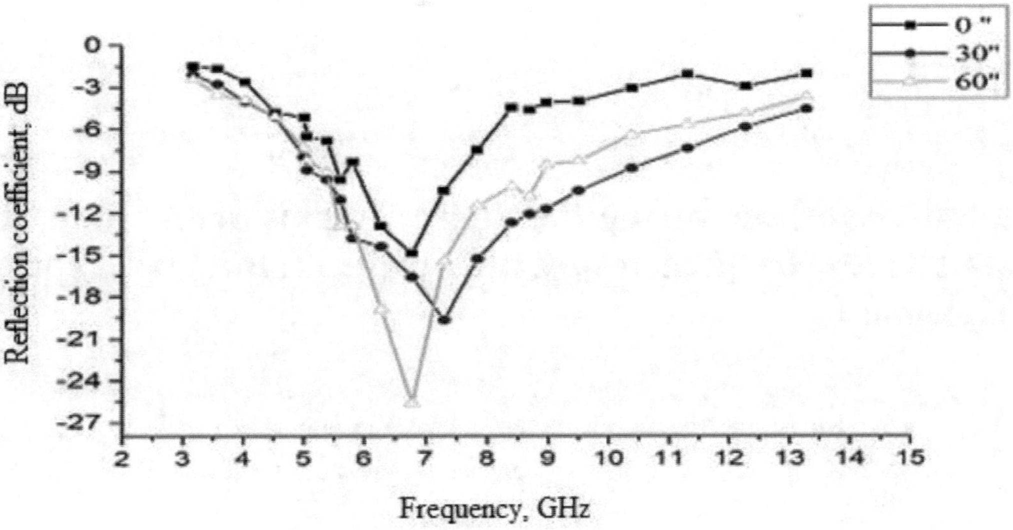

Figure 4. Frequency dependence of the reflection coefficient of nanosized powders of the hexaferrite system $Co_{0.7}Zn_{1.3}W$

3. Conclusion

This article presents the frequency dependence of the complex permeability and complex permittivity, as well as the reflection coefficient of nanosized powders of hexaferrite of the Co0.7Zn1.3W system.

Thus, it can be seen from the results obtained that nanotechnology for the production of hexaferrites makes it possible to change their dynamic characteristics, that is, to create materials with the desired properties.

4. Acknowledgement

We express our gratitude to Associate Professor of the Department of Radio Electronics Suslyaev V.I. for help in the discussion of the results.

References

[1] Zhuravlev V.A., Naiden E.P., Suslyaev V.I., Korovin E.Yu., Minin R.V., Itin V.I. 2015 IOP Conference Series: Materials Science and Engineering **81** 012003

[2] Muratov D.G., Kozhitov L.V., Korovushkin V.V., Korovin E.Y., Popkova A.V., Novotortsev V.M 2019 Russian Physics Journal **61** 10.

[3] Moseenkov S.I., Kuznecov V.L., Zavorin A.V., Golubcov G.V., Korovin E.YU., Suslyaev V.I., Ishchenko A.V., Serkova A.N., Sergeenko D.I., Velikanov D.A. 2020 ZHurnal prikladnoj himii **93** 4.

[4] Najden E.P., ZHuravlev V.A., Minin R.V., Itin V.I., Korovin E.YU. 2015 Izvestiya vysshih uchebnyh zavedenij. Fizika **58** 1.

[5] Zhuravlev V.A., Suslyaev V.I., Naiden E.P., Kirichenko V.I. 1990 Soviet Physics Journal 9 107.

[6] Moseenkov S.I., Kuznetsov V.L., Zavorin A.V., Golubtsov G.V., Korovin E.Y., Suslyaev V.I., Ishchenko A.V., Serkova A.N., Sergeenko D.I., Velikanov D.A. 2020 *Russian Journal of Applied Chemistry* **93** 4.

[7] Korovin E.Yu., Suslyaev V.I., Ul'yanova O.A. Svidetel'stvo o registracii programmy dlya EVM RU 2017611876 2017.

[8] Korovin E.Yu., Suslyaev V.I., Ul'yanova O.A Svidetel'stvo o registracii programmy dlya EVM RU 2017611446 2017.

IOP Conf. Series: Materials Science and Engineering 1198 (2021) 012006 doi:10.1088/1757-899X/1198/1/012006

Installation for measuring the dielectric anisotropy of liquid crystals at low frequencies by the bridge method with constant displacement

S V Kalashnikov[1,2*], N A Romanov and A V Nomoev[1,2]

[1]Dorzhi Banzarov Buryat State University, 24a Smolina str., Ulan-Ude, 670000, Russia

[2]Institute of Physical Materials Science SB RAS, 6 Sakhyanova str., Ulan-Ude, 670047, Russia

e-mail: Betch_kail@mail.ru

Abstract. Installation designed to measure the dielectric anisotropy in laboratory studies of liquid crystal polymer films is described. The installation operates on the principle of a balanced alternating current (AC) bridge, allowing the application of a direct external current (bias) to the liquid crystal cell. The internal resistance of the direct current (DC) source, which affects the equilibrium condition of the bridge, is compensated. The frequency of the AC current feeding the bridge and the offset voltage of the cell is regulated within a wide range, which makes it possible to study various functional dependences of the dielectric parameters of liquid crystals and their modifiers.Introduction

1. Introduction

The theoretical and applied significance of liquid crystals is well known. An external influence, for example, an electric field, causes a change in some of its physical properties. Therefore, liquid crystals (LC) have found wide application as an anisotropic medium in scientific research, information display devices, optical-control devices, and even microwave technology [1]. Nowadays, many LC is known and synthesized, modification of LC films with various micro and nanoparticles is developing [2], which makes it possible to improve the properties of LC, for example, their speed.

LC, being dielectrics, are polarized when an electric field is applied to them; the reorientation of molecules leads to a change in their dielectric constant. Therefore, the dielectric anisotropy of LC is an important characteristic. The possibility of investigating the functional dependence of the dielectric anisotropy of LC on the strength of the electric field applied to them and on the frequency of the alternating current, using which the cell impedance is measured, makes it possible to subtly study LC and their interaction with various modifiers that improve their properties. For example, there is a hysteresis of the dielectric anisotropy (and, consequently, the cell capacity) of LC from the applied voltage, modified by some nanoparticles [4], making it possible to create liquid crystal cells with particular properties.

There are various methods for measuring the dielectric anisotropy of LC or their films: resonance, bridge, beat methods, determination of the time constant of an RC circuit, and various versions of microwave methods [5]. Often, however, it is important to measure the dielectric anisotropy of LC in the absence of high-frequency currents or high-frequency electromagnetic radiation. All methods for measuring dielectric anisotropy at low frequencies or direct current are reduced to consider the change in the cell capacitance of LC with different orientations of their molecules. The direction of molecules is usually set by using orienting layers of various configurations in a liquid crystal cell [6]. However, this method has significant drawbacks; for example, it is impossible to quickly measure the parameters of its optical speed and its dielectric anisotropy for the same cell, which reduces the statistical volume of experimental data and the reliability of research. Forced reorientation of liquid crystal molecules by applying a constant electric field when measuring the cell capacitance followed by calculating the dielectric constant based on the formula for a flat capacitor is inapplicable to most methods for measuring capacitance. So, the most common modern impedance meters are based on determining the time constant of an RC circuit - the time required for the voltage across a capacitor to reach 63.2% of its voltage when fully charged. Applying an external voltage to the measured cell will lead to the fundamental impossibility of measuring capacitance with this method.

The most suitable method for measuring the capacitance of a cell of LC at different orientations of their molecules due to the electric field created by applying a constant voltage to the cell is an AC bridge. This method has high accuracy, is insensitive to a continuous bias on one of its arms, and allows you to combine the measurement of dielectric anisotropy and opto-temporal properties of LC in one setup.

2. Brief theoretical remarks

The determination of the capacitance by an AC bridge is based, on the measurement of the electrical impedance of the measured two-terminal, in our case, a liquid crystal cell. Size of the impedance of LC, however, and liquid dielectrics or electrolytes, in general, are almost always associated with the need to separate the impedance components [7]. In polar dielectrics, which fully include LC, one can note the impedance of the volume of their layer and the impedance of a thin layer near the electrodes of the liquid crystal cell. The volume impedance is determined not only by the oriented molecules of LC but also disadvantageous, by a contribution to it by the conductivity of the liquid crystal layer itself or their film. The latter depends on the presence and concentration of ions in the coating, their charge and mobility [7]. The near-electrode impedance is mainly associated with forming an electric double layer at the surface of the cell electrodes.

For a complete description of the electrical properties of a liquid crystal cell, you can use an equivalent electrical circuit, which, however, in our case, neglects the capacity of the electric double layer (figure 1). $R_{el.}$ is the resistance of the cell electrodes; this also includes the resistance of the contacts of the sample holder and other circuit elements up to the capacitance meter; C_{LQ} – capacitance of a cell with a dielectric in the form of LC; R_{LQ} is the volume resistance of LC in the cell gap and $R_{tr.}$ is the active resistance to the current flowing due to the transfer of charges from the external electrical circuit through the electrode to the liquid crystal layer and the extraction of leads from this layer through the electrode back to the outer course. The resistance $R_{tr.}$, the active and reactance of the electric double layer are dielectric losses; they are responsible for the imaginary part of the dielectric constant.

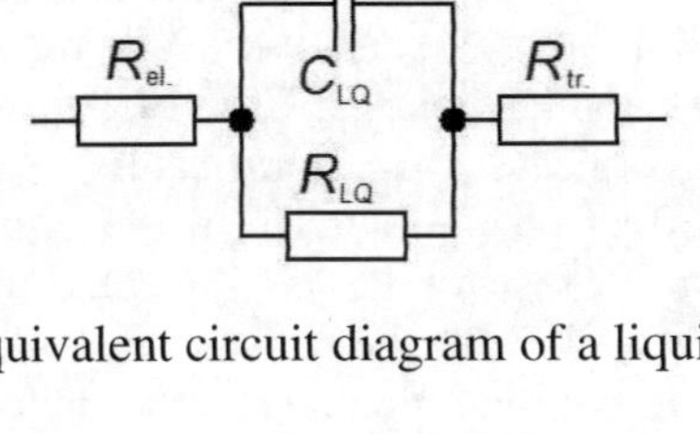

Figure 1. Equivalent circuit diagram of a liquid crystal cell.

As a result, by measuring the capacitance C of the cell by the AC bridge, it is possible to calculate the real part of the dielectric constant ε 'of LC using the well-known formula:

$$\varepsilon' = \frac{Cd}{\varepsilon_0 S},$$

where d is the thickness of the liquid crystal layer in the cell, S is the area of one of its electrodes.

To calculate the imaginary part of the dielectric constant ε", it is necessary to measure the tangent of the dielectric loss angle [6]:

$$\varepsilon'' = \varepsilon' \text{tg}\delta,$$

Thus, an AC bridge, allowing the measurement of cell capacitance, dielectric loss tangent and at the same time insensitive to the DC bias, is best suited for determining the dielectric constant of LC.

1. Device and principle of operation

A balanced AC Schering bridge was chosen for the study of LC. Unbalanced bridges, although they are more convenient to use without additional automation tools and allow a direct report from the indicator, are still less accurate [8].

A simplified diagram of an AC bridge is shown in Figure 2. The bridge is powered by alternating current from the Gen 1 generator through the TV1 transformer. The generator allows you to change the frequency of the current over a wide range. Directly the bridge arms are composed of elements R1, R2 and R4, C2 and a CL1 liquid crystal cell. The balancing of the bridge is performed by the variable resistor R1. The reference capacitor of bridge C2 is precision. The variable resistor R4 connected in series allows balancing the bridge relative to the dielectric losses in the liquid crystal cell by its position. Thus, the tangent of the dielectric loss angle can be read.

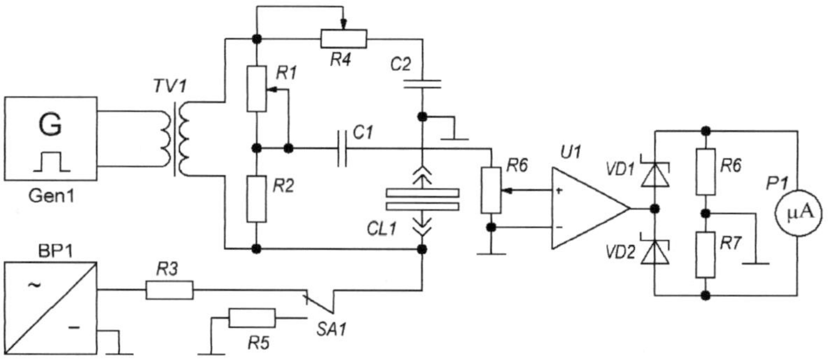

Figure 2. Simplified circuit diagram of a DC-biased AC bridge.

Like many liquid dielectrics, even well-purified LC have a noticeable ionic conductivity, the existence of which is undesirable, since the presence of a current through the cell, on the one hand, leads to the accumulation of electric charges, on the other hand, to their relaxation near the electrodes, which can worsen both the quality of the cell and the accuracy of measuring its capacity.

Amplifier U1, rectifier VD1-VD2-R6-R7 and microammeter P1 are used as a zero indicator, according to the readings of which the bridge is balanced. Using a two-half-period rectifier based on Schottky diodes with the replacement of two arms with resistors allows minimizing the voltage drop and, accordingly, increasing the null indicator's sensitivity. When measuring the parallel component of the dielectric constant, a DC bias is applied to the LC cell CL1 from a variable current source BP1 through a resistor R3. This resistor is necessary to significantly increase the internal resistance of the DC source so that it has a minor effect on the balance of the bridge. When measuring the standard component of the dielectric constant, a resistor R5 is connected in parallel to the liquid crystal cell instead of the current source by the switch SA1, the resistance of which is equal to the source's internal resistance R3. Thus, when measuring both components of the dielectric constant, the exact resistance is connected in parallel to the cell, which avoids distortions in the bridge reading when switching measurement modes. The capacitor C1 prevents the zero-indicator of the constant bias from reaching the input of the amplifier; the variable resistor R6 adjusts the amplifier's sensitivity. The amplifier is powered by a separate current source to eliminate parasitic circuits.

The equilibrium condition of the considered bridge circuit can be expressed as follows:

$$\frac{1/\omega C_{L1}}{1/\omega C_2}=\frac{R_1}{R_2}, \qquad \frac{\dfrac{R_{3,5}}{\omega C_{L1}R_{3,5}+1}}{1/\omega C_2}=\frac{R_1}{R_2} \Rightarrow \frac{\omega C_2 R_{3,5}}{\omega C_{L1}R_{3,5}+1}=\frac{R_2}{R_1},$$

$$C_{L1}=C_2\frac{R_1}{R_2}-\frac{1}{\omega R_{3,5}}, \qquad (1)$$

where $R_{3,5}$ is the resistance of the resistor *R3* or *R5* connected in parallel with the measured cell. From formula (1) it can be seen that the capacity of a liquid crystal cell in the case of its parallelization by an active circuit with a finite resistance depends on the frequency of the current at which the bridge operates. Therefore, if the bridge's scale is linearly graduated, then when measuring the dependence of the cell capacitance on frequency, an error will occur, determined by the second term of formula (1). Graphically, the value of this deviation in determining the capacitance ΔC for a natural bridge, depending on the value of the frequency of its power supply, is shown in Figure 3. Only in a separate frequency range (from ~ 10 kHz) can this dependence be considered linear and relatively insignificant (based on the capacity of a typical laboratory liquid crystal cell, the error is no more than 2-5%).

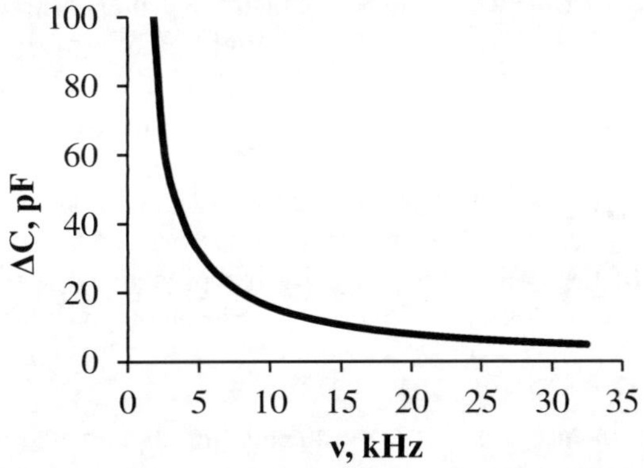

Figure 3. Dependence of the error in determining the capacitance ΔC on the current frequency v.

The tangent of the dielectric loss angle in the liquid crystal cell based on the bridge circuit can be determined as follows:

$$\operatorname{tg}\delta=\omega C_2 R_4.$$

In an actual design, the bridge is combined with equipment for measuring the optical and temporal properties of LC in electric and magnetic fields (the external view of the setup is shown in Figure 4), which makes it possible to measure various parameters for the same cell sample quickly. The report of the capacitance in the AC bridge is carried out from a vernier connected to a variable multiturn resistor R3 and representing two drums with scales, one of which, serving for a rough reading, rotates due to the planetary gear ten times slower than the second, which helps for accurate reading (Figure 5). The SA1 switch is made in the form of an electronic trigger controlled by a push-button controller and having a light indication of the bridge operation mode.

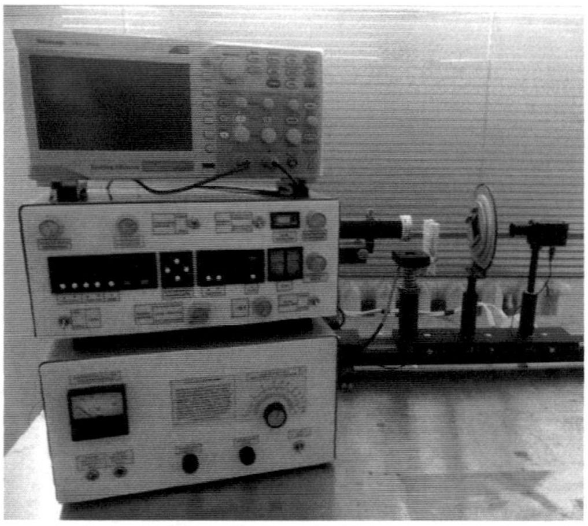

Figure 4. External view of the installation for the study of LC.

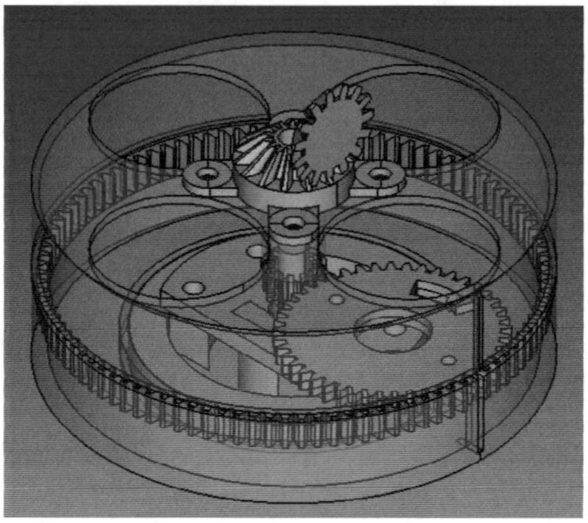

Figure 5. Three-dimensional model of the bridge vernier: at the bottom - the drum for rough reading of the container, at the top - for the accurate one (the drums are shown as translucent for clarity).

The bridge is calibrated taking into account the active resistance and capacitance of the lead wires and contacts to the liquid crystal cell using exemplary capacitors.

The measurement limits of the cell capacitance by the presented device are 100-500 pF; the loss tangent is up to 0.05, the maximum applied constant electric displacement to the cell is up to 320 V.

2. Conclusion

The above design of the installation for calculating the dielectric anisotropy by measuring the capacitance of a cell of LC meets the tasks set and can be used in other experimental studies in laboratory practice. The installation is quite simple, it uses the available elements, which simplifies its manufacture and requirements for the experimental conditions.

Acknowledgments

This work was supported by the Russian Scientific Foundation [project number 18-79-10143].

References

[1] Maune H, Jost M, Reese R, Polat E, Nickel M and Jakoby R 2018 *Crystals.* **8** 355
[2] Kim D, Kim K, Kim H, Choi M and Na J-H 2021 *Materials.* **14** 932
[3] Kalashnikov S V, Romanov N A and Nomoev A V 2016 *J. Appl. Phys.* **119** 094304
[4] Gdovinova V, Tomasovicova N, Jeng S-C, Zakutanska K, Kula P and Kopcansky P 2019 *J. Molecular Liquids* **282** 286
[5] Nayek P, Ghosh S, Karan S, Majumder T P and Roy S K 2008 *Appl. Phys. Lett.* **93** 112905
[6] Kamanina N 2019 *CBU International Conference Proceedings* **7** 933
[7] Geoffrey Luckhurst R 2015 *Biaxial nematic liquid crystals. Theory, simulation and experiment* (Chennai: John Wiley & Sons Limited)
[8] Johnson D H 2016 *Fundamentals of Electrical Engineering I* (Houston: Rice University)

New Nanomaterials and their Electromagnetic Properties (NNEP 2021) IOP Publishing
IOP Conf. Series: Materials Science and Engineering 1198 (2021) 012007 doi:10.1088/1757-899X/1198/1/012007

Towards structured SPP manipulation of light at the nanoscale

I V Minin[1], C-Y Liu[2] and O V Minin[1]
[1]School of Nondestructive Testing, Tomsk Polytechnical University, Tomsk, Russia
[2]Department of Biomedical Engineering, National Yang Ming Chiao Tung University, Taipei City, Taiwan
E-mail: prof.minin@gmail.com

Abstract. Surface plasmon photonics is a rapidly developing area of physics, optics, and nanotechnology. The unique ability of meso- and nano-structures to manipulate light in the subwavelength range down to nanoscale volumes stimulated their use in a vast research endeavours. The investigations are driven by interests in both fundamental and practical applications aspects where plasmonic light concentrators elegantly interface mesoscale dielectric structure with thin metal films. The effects of a photonic nanojet and a photonic hook, discovered by Minins, have been studied in sufficient detail in the literature, but only recently have they been able to be confirmed experimentally for low-dimensional systems – in-plane surface plasmon waves. The nature of these phenomenas lies in the dispersion of the phase velocity of waves inside the dielectric structure, which leads to constructive interference of the transmitted, diffracted, and near-field waves. Our results set the grounds for in-plane plasmonic wavelength scaled optics with unprecedented control of the energy flow at the nanoscale, and shown a way toward realizing the densely packed optical elements needed for future plasmonic and optical devices.

1. Introduction

The manipulation of subwavelength-scaled optical fields is attracting great interest owing to the key potential applications, including ultrafast communication, integrated plasmonic circuits, biophotonics, molecule diagnostics and quantum optics on a chip [1-5]. To this end the surface plasmon polaritons (SPPs) are perspective candidates for studying different optical phenomena and manipulate light at the nanoscale because SPPs are highly localized at metal–dielectric interface decaying exponentially away from it and thus have a shorter wavelength than the excitation radiation [6]. Such a planar system with a subwavelength confinement is attractive for "flatland" optics and photonics [7]. Different structures based on curved slits, circular and rectangle holes, dielectric structures, parabolic mirrors and dielectric metasurfaces have been proposed to realize subwavelength focusing [8-16] as the localization and propagation properties of SPPs are dictated by the surface morphology of the interface between the dielectric and metal film.

2. Plasmonic nanojet

The theoretical possibility of SPP nanojet (SPPJ) formation based on simulations was considered in [17-20] for dielectric structures of different shape. It has been shown that the effective refractive index contrast lies between 1.3 and 1.75. It was also demonstrated by the simulations that it is possible to produce SPPJs using dielectric mesoscale particles with a subwavelength focal spot and detect subwavelength metal particles [18]. The experimental observation of SPPJ propagation for the SPP waves under the optical excitation of telecom λ_0 = 1530 nm are shown in Figure 1 [21]. Structures were manufacturing at Technical University of Denmark. The experimental value of FWHM (Full Width at Half Maximum) was equal to 0.68 λ_0 and the SPPJ propagation length of 3.53 λ_0.

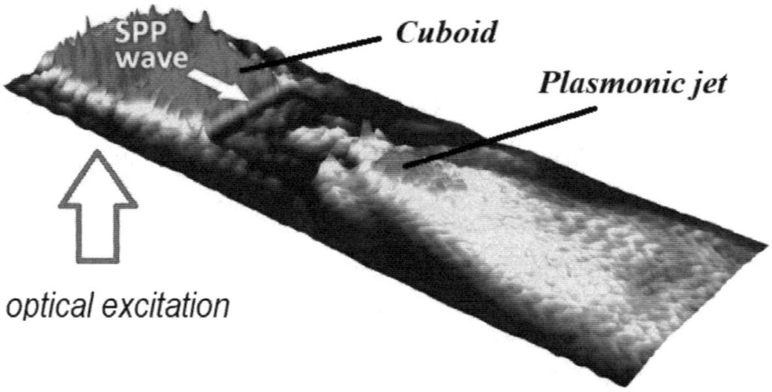

Figure 1. Plasmonic nanojet propagation for the SPP waves. The optical excitation of telecom wavelength λ_0 = 1530 nm is incident on the backside of the diffraction grating resulting in the excitation of SPP waves. Adapted from [22].

3. Plasmonic hook

Modern in-plane optical applications require so-called structured beam, tailored in shape (trajectory), amplitude and phase. It is well known how to spatially structure light [23], first demonstrated by Young in fringes about 200 years ago and more recently for SPP, in so-called structured beam [24-26]. SPP Airy beams, as example of structured beams, have been suggested theoretically [27] and confirmed experimentally [28,29] in plasmonics. Until recently, in low-dimensional systems SPP waves propagating along the curved trajectory were only SPP Airy-family beams.

A new class of curved localized surface plasmon wave, called the SPP hook (SPPH), was theoretically introduced in [30] and was suggested on the basis of in-plane manipulating SPPs as classical light in 3D free space, discovered in [31] and investigated in [32-39]. When we shift from 3D free space into two dimensions (in-plane), the quantum nature of matter manifests itself, creating new options for manipulation of light at nanoscale and controlling the interaction between matter and light [40]. For SPP waves we use of dielectric thin-film structures of controlled geometric shapes with broken symmetry on top of a metal (Gold) film surface [41]. To obtain the intensity distribution of the SPP wave field near the dielectric structure, the method of scanning near-field optical microscopy was used [42,43].

The SPPH is created using the in-plane focusing of the SPP wave through a dielectric Janus particle (broken symmetry of the shape). Mesoscale dielectric particle is a simple planar structure overcoming several of limited functionality and difficulty of integration. The word "particle" in this content means a mesoscale dielectric particle, i.e. the size of which is larger and comparable to the radiation wavelength, when strong coupling of scattered and evanescent waves in the near-field region is important [31]. Evidently, it will play a key role in the development of the future research of miniature and high-precision equipment. The SPPH propagates along wavelength scaled curved trajectory with radius less than the SPP wavelength, which represents the smallest curvature of the beam ever recorded for the SPP, and can exist despite the strong energy dissipation at gold film surface. While being by nature non-paraxial beams, SPPH demonstrates a number of remarkable properties similar to their three-dimensional analogue in free space. In addition, SPPH tightly confines optical energy near the metal-air interface and demonstrates significant beam bending over their propagation.

Recently, we have reported on the first experimental observation of a SPP hook, which is an in-plane localized SPP beam self-bending at the sub-wavelength scale [44,45] (see Figure 2). We use a dielectric AR-P 6200 resist [45] that was placed on a 100 nm thick Au film. The dielectric Janus particle has widths of cuboid rib of about from 4000 to 5100 nm, with thickness from 220 to 280 nm and the angle of the prism is equal to 27°. The samples were prepared in Department of Photonics Engineering, Technical University of Denmark. It is obvious from the Figure 2 that there is an inflection point in the SPPH, where the SPP beam changes the direction of propagation [31,44] at a distance from the rear surface about the SPP wavelength. This property shows one of the key differences between well-known SPP Airy-like beams and SPPH [46].

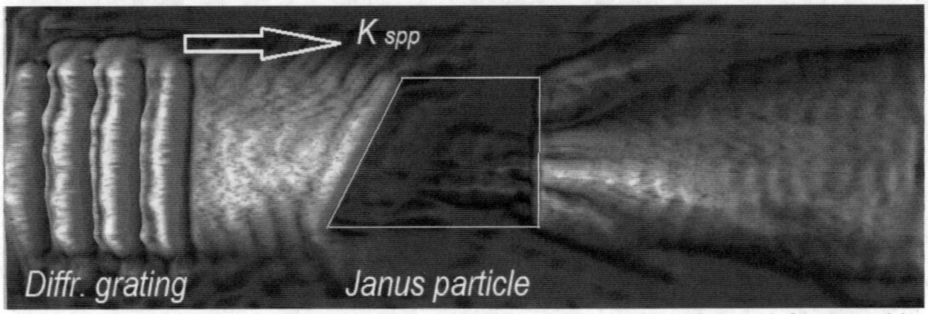

Figure 2. Experimental images of the SPPH. The diffraction grating is used for launching the SPP wave [47]. The SPP wave with a wave vector k_{spp} is excited at the telecom wavelength of λ_0= 1530 nm and is incident on the SPPH from left to right. The FWHM and propagation length are 1180 nm and 7660 nm, respectively. The beam-bending angle was equal to 15 degree for structure with width of 5 nm. Adapted from [44].

The reason for the appearance of the bent form of the localized SPP beam is constructive interference between the incident, diffracted, and scattered fields near the shadow surface of the dielectric structure due to interactions in the near field. The bending effect of the SPP beam (curvature) at a fixed shape of the dielectric structure depends on its size. A decrease in size leads to a decrease in the effective length of the SPP hook and full propagation lengths, but a slight increase in its curvature (and the SPPH deflection angle) due to the stronger manifestation of diffraction effects on the edges of the structure. So, for the structure shown in Figure 2, the change in the bending angle is in the range from 3 to 16 degrees with a change in wavelength by $\pm$2% with a maximum corresponding to zero detuning. At the same time, with a decrease in the size of the structure, the change in the bending angle is from 6 to 18 degrees with a maximum corresponding to a change in the wavelength by +2%, i.e. moves to the right. This is an additional degree of freedom that can be used to control light to a nanoscale level and dynamical manipulation of SPP waves including switching. The optically controlled electronic circuitry for energy-efficient and faster switching with photonic functionality is an appealing notion.

4. Discussions

Well-known idiom since ancient times is ''Seeing is believing''. The concepts of photonic nanojet and hook now has gone beyond classical optics and penetrated such fields, as THz, acoustic [48-50] and now low dimension system - surface plasmons. Highly confined structured low-dimension electromagnetic fields play a significant role in SPPs because of their outstanding subwavelength nature [51,52]. The plasmonic jet and hook represents an exciting new field for the application of surface-plasmon waves in which SPP based concepts merge the fields of electronics and photonics at the nanoscale [3,53]. We succeed in developing the first in-plane plasmonic wavelength-scaled refractive-diffractive lens. The results described above open the exciting possibility of a readily available structured SPP beam toolkit that can both create and deliver exotic states of beam at nanoscale. It is remarkable that this possibilities is achieved with inexpensive and a simple unusual optical element.

It is interesting to note that recently local excitation of plasmonic PMMA waveguides by hole was considered in [54]. The results are shown in Figure 3. The structure consists of a circular hole with 800 nm diameter milled in 200 nm thick gold film. The hole is adjacent to a 5 μm long and 300 nm thick PMMA line. The field confinement 0.5 μm after the end of the dielectric rectangle similar to SPP nanojet. However, the characteristics of the localization region of the SPP in the shadow part of the dielectric particle are not given.

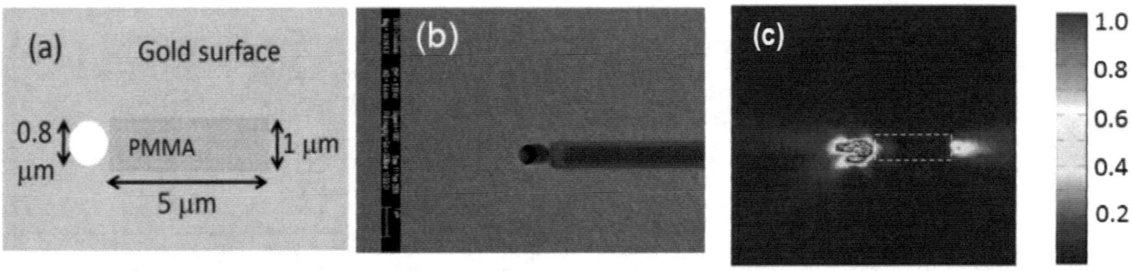

Figure 3. (a) Proposed in [54] structure. (b) Fabricated sample. (c) NSOM measurement results. Adapted from [54] under a Creative Commons Attribution 4.0 International License.

From a future applications possibilities, photonic and plasmonic hooks are important because the Airy-family beams now are not the only self-bending beams. Importantly, these effects prove that the concept of wave acceleration can be observed in non-paraxial approximation in the near-field due to constructive interference of refracted and diffracted waves. And investigations shown that photonic and plasmonic nanojet and hook are exceptionally easy to manipulate in unusual ways and realize. The effects of the SPP nanojet and hook could open new horizons for manipulating the interaction between matter and light at the nanoscale. One may highlight several key potential applications of mentioned above effects: plasmonic trapping and tweezers [55,56], plasmonic and nanoscale biosensing sensors and nature [57-60]. These effects also could be potentially used to increase photocarrier confinement in THz receivers, emitters and antennas [61-66].

5. Conclusions

Full two-dimensional in-plane control of structured light through the use of near-field effects and non-paraxial approaches based on common dielectric materials may open a new possibilities for control of localized beam parameters beyond the diffraction limit in printed photonic-driven devices. For example, we believe that the curved trajectory of the photonic hook structured beam can break the limitations of classical condition of line-of-sight in in-plane photonic interconnection [67,68] and switching on-a-chip. These exciting perspectives are surely worthy of further deep attention from the optical community. It is our great hope that this work should attract the attention of scientists to get engaged with plasmonic investigations and its potential applications, including on-chip virus detection [69-72] and provides new ideas for the development of new in-plane devices [73,74].

Acknowledgments

This work was partially supported by the Russian Foundation for Basic Research (Grant No. 20-57-S52001), by TPU Development Program and by Ministry of Science and Technology of Taiwan (MOST 108-2221-E-010-012-MY3, MOST 109-2923-E-010-001-MY2), and Yen Tjing Ling Medical Foundation (CI-110-28).

References

[1] Andrew P and Barnes W 2004 *Science* **306** 1002
[2] Ozbay E 2006 *Science* **311(5758)** 189–193

[3] Anger P, Bharadwaj P and Novotny L 2006 *Physical Review Letters* **96(11)** 113002
[4] Yin L, Vlasko-Vlasov V, Rydh A 2004 *Applied Physics Letters* **85(3)** 467–469
[5] Aoni R, Rahmani M, Xu L, Kamali K, Komar A, Yan J, Neshev D and Miroshnichenko A 2019 *Sci Rep* **9** 6510
[6] Raether H 1988 *Surface Plasmons on Smooth and Rough Surfaces and on Gratings* (Springer-Verlag, Berlin).
[7] Chen P, Argyropoulos C, Farhat M and Gomez-Diaz J 2017 *Nanophotonic* **6** 1239
[8] Kim H, Cheng X 2011 *Opt. Lett.* **36(16)** 3082–3084
[9] Wu G, Chen J, Zhang R, Xiao J and Gong Q 2013 *Opt. Lett.* **38(19)** 3776–3779
[10] Mote R, Minin O V and Minin I V 2017 *Opt. Quant. Electron.* **49** 271
[11] Shi W B, Chen T Y, Jing H, Peng R W and Wang M 2017 *Opt. Express* **25(5)** 5772-5780
[12] Long Y, Zhang Z, Su X 2016 *EPL* **116** 37004
[13] Melentiev P, Kuzin A, Negrov D and Balykin V 2018 *Plasmonics* **13(6)** 2361–2367
[14] Yin X, Steinle T, Huang L, Taubner T, Wuttig M, Zentgraf T and Giessen H 2017 *Light: Science & Applications* **6** e17016
[15] Gartcia-Ortiz C, Cortes R, Gomez-Correa J, Pisano E, Fiutowski J, Garcia-Ortiz D, Ruitz-Cortes V, Rubahn H G and Coello V 2019 *Photonics Research* **7(10)** 1112
[16] Cherukulappurath S, Heinis D, Cesario J, van Hulst N, Enoch S and Quidant R 2009 *Opt. Express* **17** 23772-23784
[17] Ju D, Pei H, Jiang Y and Sun X 2013 *Appl. Phys. Lett.* **102** 171109
[18] Pacheco-Pena V, Minin I V, Minin O V and Beruete M 2016 *Ann. Phys.* **528** 1–9
[19] Minin I V, Minin O V, Pacheco-Peña V and Beruete M 2015 *Appl. Phys. Lett.* **106** 254102
[20] Pacheco-Pena V, Minin I V, Minin O V and Beruete M 2016 *Photonics* **3** 1–7
[21] Minin I V, Minin O V, Glinskiy I, Kabibullin R, Malureanu R, Lavrinenko A V, Yakubovsky D, Volcov V and Ponomarev D 2020 *Optics Letters* **45(12)** 3244-3247
[22] Minin I V, Minin O V, Glinskiy I, Yakubovsky D, Volcov V and Ponomarev D 2019 *arXiv*: 1912.13373
[23] Vijayakumar A, Rosales-Guzman C, Rai M, Rosen J, Minin O V, Minin I V and Forbes A 2019 *Opt. express* **27(5)** 6459-6470
[24] Ravets S, Rodier J, Kim B, Hugonin J, Jacubowiez L and Lalanne P 2009 *J. Opt. Soc. Am. B* **26** B28-B33
[25] Aberra Guebrou A, Laverdant J, Symonds C, Vignoli S and Bellessa J 2012 *Opt. Lett.* **37** 2139-2141
[26] Hu Z, Tan P, Zhu S and Yuan X 2010 *Opt. Express* **18(10)** 10864
[27] Salandrino A and Christodoulides D N 2010 *Opt.Lett.* **35** 2082
[28] Minovich A, Klein A, Janunts N, Pertsch T, Neshev D and Kivshar Y 2011 *Phys. Rev. Lett.* **107** 116802
[29] Li L, Li T, Wang S, Zhang C and Zhu S 2011 *Phys. Rev. Lett.* **107** 126804
[30] Minin I V, Minin O V, Ponomarev D and Glinskiy I 2018 *Ann.Phys.* **530(12)** 1800359
[31] I. V. Minin and O. V. Minin 2016 *Diffractive Optics and Nanophotonics: Resolution below the Diffraction Limit* (Cham: Springer).
[32] Yue L, Minin O V, Wang Z, Monks J, Snalin A and Minin I V 2018 *Opt. Lett.* **43** 771
[33] Ang A, Karabchevsky A, Minin I V, Minin O V, Sukhov S and Snalin A 2018 *Scientific Reports* **8** 2029
[34] Minin I V, Minin O V, Katyba G, Chernomyrdin N, Kurlov V, Zaytsev K, Yue L, Wang Z and Christodoulides D N 2019 *APL* **114(3)** 031105
[35] Minin I V, Minin O V, Liu C, Wei H, Geints Y and Karabchevsky A 2020 *Optics Letters* **45(17)** 4899-4902
[36] Spector M, Ang A, Minin O V, Minin I V and Karabchevsky A 2020 *Nanoscale Advances* **2** 5312-5318
[37] Minin I V, Minin O V, Liu Y, Tuchin V and Liu C 2021 *J Biophotonics* **14(2)** e202000342

[38] Minin I V, Minin O V, Zaytsev K, Katyba G, Kurlov V, Yue L, and Wang Z 2020 *Proc. SPIE* 1136807 (2 April 2020)
[39] Minin I V and Minin O V 2021 *IOP Conf. Ser.: Mater. Sci. Eng.* **1019** 012093
[40] Novotny L and Hecht B 2021 *Principles of Nano-Optics, 2nd ed.* (Cambridge University Press).
[41] Burke J, Stegeman G and Tamir T 1986 *Phys. Rev. B* **33** 5186-5201
[42] Huber A J, Deutsch B, Novotny L and Hillenbrand R 2008 *Appl Phys Lett* **92** 203104
[43] Alfaro-Mozaz F, Alonso-González P, Vélez S, Dolado I, Autore A, Mastel S, Casanova A, Hueso L, Li P, Nikitin A and Hillenbrand R 2017 *Nature Communications* **8** 15624
[44] Minin I V, Minin O V, Glinskiy I, Kabibullin R, Malureanu R, Yakubovsky D, Volcov V and Ponomarev D 2020 *ArXiv*: 2004.10749
[45] Minin I V, Minin O V, Glinskiy I, Kabibullin R, Malureanu R, Lavrinenko A V, Yakubovsky D, Volcov V and Ponomarev D 2021 *APL* **118(13)** 131107
[46] Minin I V and Minin O V 2020 *J. Biomed. Photonics Eng.* **6(2)** 020301
[47] Koev S T, Agrawal A, Lezec H J and Aksyuk A A 2012 *Plasmonics* **7** 269–277
[48] Minin O V and Minin I V 2017 *Opt. Quantum Electron.* **49(2)** 54
[49] Rubio C, Tarrazó-Serrano D, Minin O V, Uris A and Minin I V 2020 *Results in Physics* **16** 102921
[50] Castiñeira-Ibáñez S, Tarrazó-Serrano D, Uris A, Rubio C, Minin O V and Minin I V 2020 *Sci Rep* **10** 20279
[51] Gramotnev D K and Bozhevolnyi S I 2010 *Nature Photon* **4** 83 –91
[52] Yang A, Meng F, Shi P, Du L and Yuan X 2019 *Opt. Express* **27** 18980-18987
[53] Law S, Podolskiy V and Wasserman D 2013 *Nanophotonics* **2(2)** 103–130
[54] Alam M Z, Yang Z, Sheik-Bahae M, Aitchison J S and Mojahedi M 2021 *Sci Rep* **11** 3170
[55] Han X and Sun C 2019 *Appl. Sci.* **9** 3596
[56] Zhang Y, Min C, Dou X, Wang X, Urbach H P, Somekh M and Yuan X 2021 *Light Sci Appl* **10** 59
[57] Tong L, Wei H, Zhang S and Xu H 2014 *Sensors* **14(5)** 7959-7973
[58] Wu F, Thomas P A, Kravets V G *et al.* 2019 *Sci Rep* **9** 20286
[59] Shrivastav A M, Cvelbar U and Abdulhalim I 2021 *Commun Biol* **4** 70
[60] Tommasi E D, Esposito E,·Romano S, Crescitelli A, Meo A D, Mocella V, Zito G and Rendina I 2021 *La Rivista del Nuovo Cimento* DOI: 10.1007/s40766-021-00015-w
[61] Minin I V, Minin O V, Delgado-Notario J, Calvo-Gallego J, Velazquez-Perez P, Ferrando-Bataller F and Meziani Y M 2020 *Phys. Status Solidi RRL* 1900700
[62] Minin O V, Minin I V, Meziani Y M, Hisatake S 2020 *Opt Eng* **60(8)** 082005
[63] Caldwell J, Lindsay L, Giannini V, Vurgaftman I, Reinecke T, Maier S and Glembocki O 2015 *Nanophotonics* **4** 44–68
[64] Tong J, Suo F, Ma J, Tobing L, Qian L and Zhang D 2019 *Opto-electronics advances* **2(1)** 180026
[65] Bacon D, Madéo J and Dani K 2021 *J Optics* **23** 064001
[66] Berry C W, Wang N, Hashemi M R, Unlu M, Jarrahi M 2013 *Natt. Comm.* **4** 1622
[67] Joy S, Erementchouk M, Yu H and Mazumder P 2018 *IEEE Transactions on Communications* **67(1)** 599
[67] Panchenko E, Cadusch J, Avayu O, Ellenbogen T, James T, Gómez D and Roberts A 2018 *Adv. Mater. Technol.* **3**, 1700196
[69] Abdulhalim I, Zourob M and Lakhtakia A 2008 *Electromagnetics* **28** 214–242
[70] Lopez G A, Estevez M C, Soler M and Lechuga L M 2017 *Nanophotonics* **6(1)** 123–136
[71] Mauriz M 2020 *Sensors* **20** 4745
[72] Shrivastav A M, Cvelbar U and Abdulhalim I 2021 *Commun Biol* **4** 70
[73] Jiang L, Yin T, Dubrovkin A, Dong Z, Chen Y, Chen W, Jang J, Zhen Z 2019 *Light Sci Appl* **8** 21

[74] Kumar A, Gupta R, Manjuladevi V, Joshi A 2021 *Plasmonics* DOI 10.1007/s11468-021-01373-1

Structured plasmonic beam: in-plane manipulation of light at the nanoscale

I V Minin[1,2] and O V Minin[1,2]
[1]School of Nondestructive Testing, Tomsk Polytechnical University, Tomsk, Russia
[2]Siberian State University of Geosystems and Technologies, Novosibirsk, Russia
E-mail: prof.minin@gmail.com

Abstract. The brief review on recent approaches on the formation of a new class of subwavelength scale localized structured surface plasmon polaritons (SPP) beams is discussed. For the Janus-like particle (including the geometrically symmetric particles with different dielectrics) the morphology of the field localization area and its properties depends on the particle shape and material. Plasmonic hook (PH) beam does not propagate along straight line but instead follow curved self-bending trajectory. Wavefront analysis behind of such symmetric and asymmetric mesoscale rectangle structure reveals that the unequal phase of the transmitted plane wave results in the irregularly concave deformation of the wavefront inside the dielectric which later leads to creation of the PH. Such dielectric structures placed on metal film enable the realization of new ultracompact wavelength-selective and wavelength-scaled in-plane nanophotonic components. SPP have potential to overcome the constrains on the speed of modern digital integrated devices limitation due to the metallic interconnects and increase the operating speed of future digital circuits.

1. Introduction

A number of works have been devoted to the various physical principles and possibilities of manipulating biological samples, materials, etc. using light [1-7] since the discovery of optical traps. Both continuous (including resonant) and pulsed light illumination have played an important role in the development of this research [1, 6-11]. Note that the effects of pulsed illumination are quite complex and can significantly change the optical force. For example, heating effects can change the dielectric constant of a particle, which can significantly increase the optical power [12].

However, it is quite difficult to capture and trap Rayleigh particles well below the diffraction limit directly using conventional optical tweezers - since the nanoparticles dimensions are much smaller than a focused beam spot, and the nanoparticle volume limits the amount of light that can be used for trapping [13]. Among others tweezers, plasmonic tweezers are based on a surface plasmon (SP) wave refer to charge density oscillations and generated at the interface between a dielectric structure and a conductive film. The nanoscale confinement of SPs excited by laser light can give rise to many scientific and practical applications. SP waves are two-dimensional in nature. However, with the transition from three-dimensional to two-dimensional space, the quantum nature of matter begins to manifest itself - new opportunities are created to control the interaction between structures and light. Among the different ways of SPs manipulation, the great interest is directed at the implementation of the simple, low cost and compact method for subwavelength localization and curving the trajectory of SP wave at the nanoscale. One of the potential applications of curved surface plasmonic beams is in the fields of surface trapping of nano-objects. The optical power generated by surface plasmon in resonant mode has been proposed and investigated since 2001 [14-20]. Plasmonic tweezers were rapidly developed in the last 10 years, resulting in improved trapping parameters and their use in different area including biological. In addition, the SP waves provide further spatial resolution towards the nanoscale [21,22] and opens new horizonts in the realization of optically driven on-a-chip devices.

In this regard, a promising direction is the development of simple and effective methods for localizing SPs in the sub-diffraction volume and even below the diffraction limit. Also among the important domains in plasmonics is the study of the transport of SPs and the development of physical methods for manipulating them at the nanoscale.

2. Wavelength-selective mesoscale dielectric structures

The in-plane Fresnel zone plate to focus the SPP waves was studied in [23] and 15 times the field intensity enhancement at the focal length of 60 nm was obtained. Later, a conical in-plane nanoslit-based Fresnel zone plate was proposed to realize focusing below the diffraction limit by exciting SPPs and enabling them to couple with radiating propagation modes [6, 24]. The focal spot size was as small as 0.31 times the illumination wavelength at the focal distance of 8 wavelength. The depth of focus was 0.6 of illumination wavelength. The theoretical possibilities of plasmonic photonic jet formation for flat dielectric disc [25] and square [26] single particles were also considered, demonstrating the capability to control the subwavelength focusing properties of a plasmonic particle-lens by profiling the height of the dielectric structure at the operation wavelength in order to obtain the required effective refractive index contrast. The resolution of plasmonic jet produced by single Si_3N_4 cube was 0,40 λ at 1300 nm wavelength with field intensity enhancement of 9 dB [26]. The array of flat dielectric Si_3N_4 cuboids [27], in despite of the dissipation of SP energy while propagation is caused due to Ohmic absorption in metals, allow to increase the propagation length of SPs. Later, the effect of the plasmonic nanojet formation was experimentally verified for the first time in [28]. The experimental value of the propagation length of plasmonic nanojet of 3.53 λ at telecom wavelength of 1530 nm (Fig.1). It can be seen that the best (minimum) resolution is observed at a small negative detuning of the wavelength of the irradiating radiation.

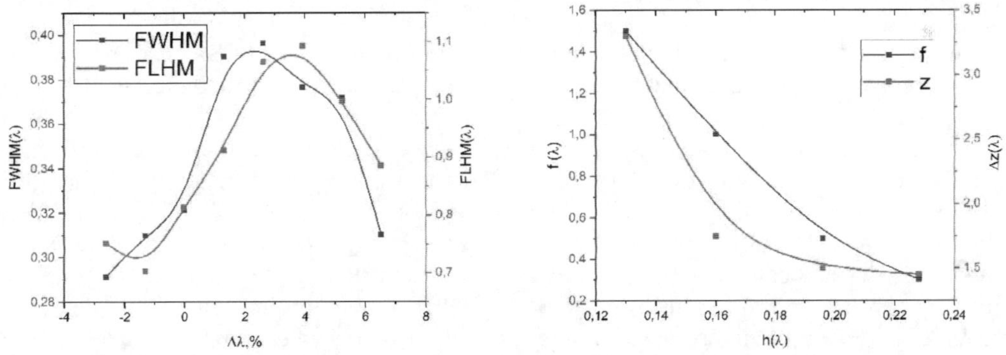

Figure 1. Dependences of the half-width of the field intensity distribution across (FWHM – full width at half maximum) and along the plasmonic jet (FLHM – full length at half maximum) vs illuminating wavelength (left) and the focal length, the length of the plasmon jet vs height of the dielectric structure (right).

For moving nanoparticles around obstacles the concept of so-called photonic hook [29] (PH) was investigated theoretically [30] and experimentally [31,32,33] in free space. At the same time, based on the discoveries in free space [29] the concept of in-plane plasmonic hook was theoretically predicted in [34] and experimentally verified for the first time in [35]. It has been shown that SP hook bend angle occurs within the excitation the telecom wavelength and propagates along a wavelength-scaled curved trajectory in contrast to that for the other today known self-bending Airy-family plasmonic beams usually having a parabolic beam trajectory. The PH angle of curvature, defined as an angle of beam bending, in the smaller dimensions structure is higher and increase with increasing of excitation wavelength than that for the more bigger structure, in which the speed of change of the curvature little with the change vs wavelength. Thus, PH could be used to focus and deflect SPs through a Janus dielectric particle [35]. Moreover, it has been shown both for electromagnetic 3D case, acoustics and in-plane SP that in the photonic hook family beams there is an inflection where the curved beam changes its propagating direction (Fig.2). This property is not possessed by the Airy-family beam and demonstrates one of the fundamental differences between subwavelength curvature photonic hook and classical Airy-like beams [6,29,31-35].

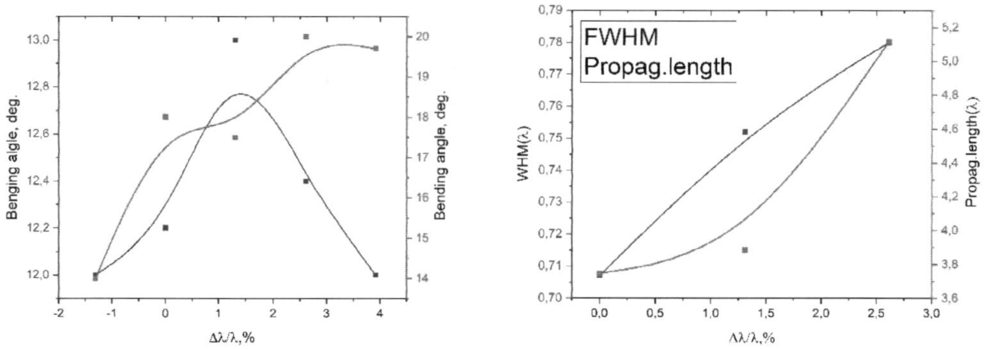

Figure 2. Bending angle (left) and propagation length and FWHM of the SPP hook (right) vs illuminating wavelength. The bending angles (left) are shown for different dimensions of dielectric Janus particle (small – red and big - black).

Our analysis for different wavelengths has shown that the situation is quite complicated. Although for a sufficiently large Janus particle size, the change in the bending angle versus frequency is not large, nevertheless, a "quasi-resonance" behavior of the considered dependence is observed. For a smaller particle, the bending is slightly larger, and the resonance structure is not visible. It is interesting to note that regardless of the size of the Janus microparticle, the range of the bending angle is the same.

3. Discussions

In summary, it was confirmed, realized and demonstrated high performance of SPP-based wavelength selective nanophotonic mesoscale elements operating at telecom wavelengths: dielectric flat cuboid plasmonic lens and plasmonic hook generator, thereby enabling the realization of ultracompact wavelength-scaled plasmonic components. Wavelength-scaled dielectric cubic-based particle is a simple planar structure overcoming the block of limited functionality and difficulty of integration. Credibly, it will play an important role in the development of the research of future miniature and high-precision equipment. The authors applied their experience in this field to solving modern SPP photonics problems. It has been shown that auxiliary dielectric structures both symmetrical and with broken symmetry can be used to modify properties of SPP manipulation in-plane, assist configuring SPP optical forces and obtain extraordinary parameters of light-matter interaction. It could be noted that in order to obtain a curved SPP flux the geometrically symmetric cubic particles, but with a specially imposed asymmetry of the refractive index (two different dielectrics) may be used. The optical manipulation of micro- and nano-objects with structured near-field SPP beams and the problem of sharp sub-wavelength structured SPP beams formation by mesoscale dielectric particles is also solved. The experiments showed that the SPPs hook propagate over 5 of wavelength at telecom wavelengths along propagation directions in gold film and exhibit strong subwavelength confinement with low bend losses [35]. Due to its flexibility, ease of fabrication and good reproducibility, the proposed platform can be adapted to accommodate future applications ranging from nanoscale light control to photonic devices on a chip. SP fields can be shaped, localized (focused) and guided by simple dielectric mesoscale structures.

4. Conclusions

The effects of localized structured field in the form of the plasmonic jet and the plasmonic hook can be used both to increase the limitation of photocarriers in terahertz photoconductive devices (PD) [36,37] by introducing single or multiple Janus dielectric particles into the PD's gap [38] and to create, for example, curvilinear waveguide structures similar to [27]. To tune the focal length and the localized structured field of surface plasmonic modes a dielectric structure based on optofluidics may be used [39,40]. Experimental results show that these in-plane refraction-diffraction methods has a general sense with which the realized plasmonic functions are expected to find promising and wide applications for future plasmonic beam technology and integrated optical circuits [19, 41-43].

Acknowledgments

This work was partially supported by the Russian Foundation for Basic Research (Grant No. 20-57-S52001) and by TPU Development Program. We are thankful to all of our collaborators of Refs. [26-28,35] for structure manufacturing at the Department of Photonics Engineering, Technical University of Denmark, and the SNOM spectral measurements.

References

[1] Ashkin A 1970 *Phys. Rev. Lett.* **24(4)** 156
[2] Phillips W D 1998 *Rev. Mod. Phys*. **70(3)** 721
[3] Moffitt J R, Chemla Y R, Smith S B and Bustamante C 2008 *Annu. Rev. Biochem*. **77(1)** 205
[4] Bradac C 2018 *Adv. Opt. Mater.* **6(12)** 1800005
[5] Gao D, Ding W, Nieto-Vesperinas M, Ding X, Rahman M, Zhang T, Lim C and Qiu C-W 2017 *Light Sci. Appl.* **6(9)** e17039
[6] Minin I V and Minin O V 2020 *J. of Biomedical Photonics & Engineering* **6(2)** 020301
[7] Minin I V, Minin O V, Cao Y, Liu Z, Geints Y and Karabchevsky A 2019 *Sci Rep* **9** 12748
[8] Ng J, Chan C T, Sheng P and Lin Z 2005 *Opt. Lett.* **30(15)** 1956
[9] Liu T-H, Chiang W-Y, Usman A and Masuhara H 2016 *J. Phys. Chem. C* **120(4)** 2392
[10] Preez-Wilkinson N D, Stilgoe A B, Alzaidi T, Rubinsztein-Dunlop H and Nieminen T A 2015 *Opt. Express* **23(6)** 7190
[11] Usman A, Chiang W-Y and Masuhara H 2012 *J. Photochem. Photobiol. A* **234** 83
[12] Spector M, Ang A, Minin O V, Minin I V and Karabchevsky A 2020 *Nanoscale Advances* **2(6)** 2595
[13] Shoji T and Tsuboi Y 2014 *J. Phys. Chem. Lett.* **5** 2957
[14] Kwak E, Onuta T, Amarie D, Potyrailo R, Stein B, Jacobson S, Schaich W, Dragnea B 2004 *J. Phys. Chem. B* **108** 13607
[15] Grigorenko A, Roberts N, Dickinson M and Zhang Y 2008 *Nat. Photonics* **2** 365
[16] Juan M L, Righini M and Quidant R 2011 *Nat. Photonics* **5(6)** 349
[17] Song Y, Han B, Chang S 2001 *Opt. Commun* **198** 7
[18] Choudhary D, Mossa A, Jadhav M and Cecconi C 2019 *Biomolecules* **9** 23
[19] Kotsifaki D and Nic Chormaic S 2019 *Nanophotonics* **8** 1021
[20] Tan H, Hu H, Huang L and Qian K 2020 *Analyst* **145** 5699
[21] Quidant R, Petrov D and Badenes G 2005 *Opt. Lett.* **30** 1009
[22] Righini M, Volpe G, Girard C, Petrov D and Quidant R 2008 *PRL* **100** 186804
[23] Fang Z, Lin C, Ma R, Huang S and Zhu X 2009 *ACSnano* **4(1)** 75
[24] Mote R, Minin O V and Minin I V 2017 *Optical and Quantum Electronics* **49(8)** 271
[25] Ju D, Pei H, Jiang Y and Sun X 2013 *Appl. Phys. Lett.* **102** 171109
[26] Pacheco-Pena V, Minin I V, Minin O V and Beruete M 2016 *Ann. der Phys* **528** 684
[27] Pacheco-Peña V, Minin IV, Minin O V and Beruete M 2016 *Photonics* **3** 10
[28] Minin I V, Minin O V, Glinskiy I, Kabibulin R, Malureanu R, Lavrinenko A, Yakubovsky D, Volcov V and Ponomarev D 2020 *Opt. Lett.* **45(12)** 3244
[29] Minin I V and Minin O V 2016 *Diffractive Optics and Nanophotonics: Resolution below the Diffraction Limit* (Cham: Springer).
[30] Yue L, Minin O V, Wang Z, Monks J, Snalin A and Minin I V 2018 *Opt. Lett.* **43** 771

[31] Minin I V, Minin O V, Katyba G, Chernomyrdin N, Kurlov V, Zaytsev K, Yue L, Wang Z and Christodoulides D N 2019 *APL* **114(3)** 031105
[32] Minin I V, Minin O V, Liu C, Wei H, Geints Y and Karabchevsky A 2020 *Optics Letters* **45(17)** 4899
[33] Minin O V, Minin I V, Zaytsev K, Katyba G, Kurlov V, Yue L and Wang Z 2020 *Proc. SPIE* 136807
[34] Minin I V, Minin O V, Ponomarev D and Glinskiy I 2018 *Ann. Phys* 1800359
[35] Minin I V, Minin O V, Glinskiy I, Knabibulin R, Malureanu R, Yakubovsky D, Volcov D and Ponomarev D 2020 *ArXiv*: 2004, 10749
[36] Yardimci N and Jarrahi M 2018 *Small* **14** 1802437
[37] Bahk Y, Park D and Kim D 2019 *Journal of Applied Physics* **126** 120901
[38] Lavrukhin D, Glinskiy I, Khabubilin R, Arsenin A, Volcov V, Minin I V, Minin O V, Zaytsev K and Ponomarev D 2019 *Proc. 8th Russia-Japan-USA-Europe Symposium on Fundamental & Applied Problems of Terahertz Devices & Technologies* Thu-1-1
[39] Zhao C, Liu Y, Zhao Y, Fang N and Huang T 2013 *Nature Commun.* **4** 2305
[40] Armin F, Shokooh-Saremi M and Mirsalehi M 2018 *IEEE Photonics Techn. Lett.* **30(5)** 427
[41] Wong H, Righini M, Gates J, Smith P, Pruneri V, and Quidant R 2011 *Appl. Phys. Lett.* **99** 061107
[42] Zeng Y, Qu S, Chen B and Chan C 2017 *Sci Rep* **7** 9959
[43] Minin O V and Minin I V 2021 *The Photonic Hook: From Optics to Acoustics and Plasmonics* (Cham: Springer)

Thermodynamic aspects of electron-beam surface modification of low-carbon steel

U L Mishigdorzhiyn[1], A S Milonov[1], P A Gulyashinov[2], N S Ulakhanov[3,1] and A P Semenov[1]

[1] Institute of Physical Materials Science SB RAS, 6 Sakhyanovoy Str., Ulan-Ude, The Republic of Buryatia, 670047, Russia
[2] Baikal Institute of Natural Management SB RAS, 6 Sakhyanovoy Str., Ulan-Ude, The Republic of Buryatia, 670047, Russia
[3] East Siberia State University of Technology and Management, 40V Kluchevskaya Str., Ulan-Ude, The Republic of Buryatia, 670013, Russia

E-mail: undrakh@ipms.bscnet.ru

Abstract. Multicomponent surface modification of carbon steels is of high interest in mechanical engineering due to its beneficial impact on machine components' and structures' surface properties. The present research was devoted to simulating the process of aluminides and borides formation on the surface of low-carbon steel during electron beam alloying and predict the phase composition of the obtained coatings. Computational thermodynamics and approximate calculation method were used to solve the problem mentioned above. Calculations were done in the temperature range between 200 and 2000 K at 10^{-3} Pa. It was discovered that the calculated and experimental data of the coating's phase composition differs significantly. The only confirmed phase that was predicted by the calculations was sodium fluoride (NaF). It was established that NaF presence in the treatment paste was redundant for the electron beam alloying because of its low reactivity in a vacuum. XRD analysis revealed the following phases in the coating: Fe_2B, Fe_3C, and $AlFe_3$.

1. Introduction

Electron-beam processing (EBP) as a method of surface modification is continuously studied and is on high demand due to its application perspectives [1-3]. It can be used to achieve protective coatings on the surface of steels and alloys in two ways: electron beam alloying (EBA) and combined treatment. The latter process comprises two stages:

- thermal-chemical surface treatment (TCT) as a first step, where diffusion layers are formed on the surface of the base metal during high-temperature exposure in treatment media in a muffle furnace [4-6];
- the second step involves modification of the obtained diffusion layers using EBP.

In both EBA and combined treatment (TCT+EBP), varying the composition or ratio of the treatment mixture components allows obtaining specific structure formation of the layers and their desired properties. For instance, surface treatment of steels and alloys with boron and aluminum results in layers with increased oxidation and wear resistance [7]. For the muffle furnace's diffusion process, the optimal ratio of B_4C: Al in treatment paste was experimentally revealed as 5:1.

Thermodynamic calculations for TCT processes as boriding and boroaluminizing are presented in papers [8,9]. In both articles, the software was used to calculate thermodynamics. Another way to predict phase composition and the probability of chemical reactions under certain conditions can be determined by Gibbs energy calculations (approximate calculation method). The chemical reactions for boroaluminizing in the furnace without a protective atmosphere are given in paper [10].

The purpose of this study is to come up with a model of synthesis of iron borides and aluminides on the surface of low-carbon steel using EBA in a vacuum. Simulation is planned to implement in two ways: using TERRA software and by approximate calculation method. These methods will be discussed more carefully below. It is also aimed to support calculations by experimental data.

2. Thermodynamic modeling

Computational thermodynamics was carried out TERRA software. The program is designed to calculate systems of arbitrary character with chemical and phase transformations. It enables the simulation of an extremely equilibrium state by implementing the method and calculation algorithm developed at Bauman Moscow State Technical University [11]. The program is linked to an extensive database of various substances' properties, making it suitable for studying compositions with arbitrary chemical composition. The study examined the reactions in the coating under the influence of an intense electron beam in a vacuum. The paste composition was taken from the atomic masses of its component substances (table 1).

Table 1. Input composition of the paste in atomic mass for calculations

Paste components	B_4C	Al	NaF
Atomic mass	56	27	42

First, reactions occurring in the coating at a temperature between range 200 and 2000 K at 10^{-3} Pa were simulated. Figure 1 shows the appearance of aluminum diboride (AlB_2), which can play a significant role in forming iron borides and aluminides. The possible rest phases were C, Na, and Na_3AlF_6.

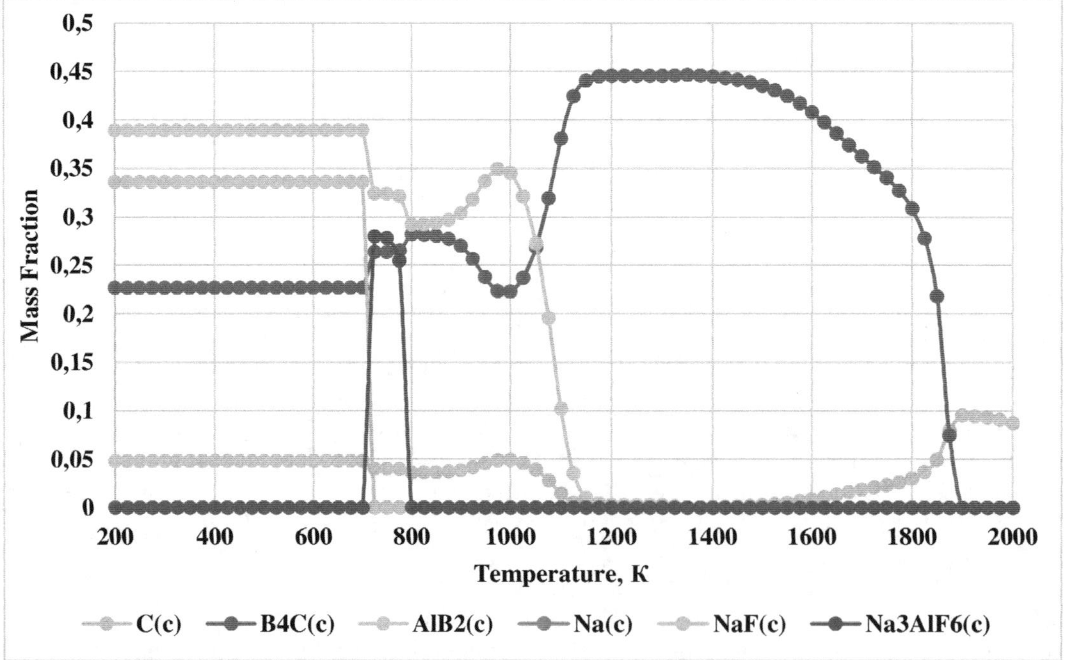

Figure 1. The output of the condensed phases in the treatment paste (T=200-2000K, P=10^{-3}Pa)

Then, the reactions between the coating and steel substrate were simulated. Low-carbon steel was used in the current research, so we neglected the steel's carbon content and other elements, except for the iron. Table 2 presents the composition of the paste and iron in the atomic masses.

Table 2. Input composition of the paste and iron in atomic mass for calculations

Paste components	Fe	B_4C	Al	NaF
Atomic mass	56	56	27	42

Figure 2 shows the output of the condensed phases during the synthesis of transition metal borides in St3 steel (T=250-2000 K, P=10^{-3} Pa). Calculations showed that iron boride FeB appeared in addition to the phases mentioned above.

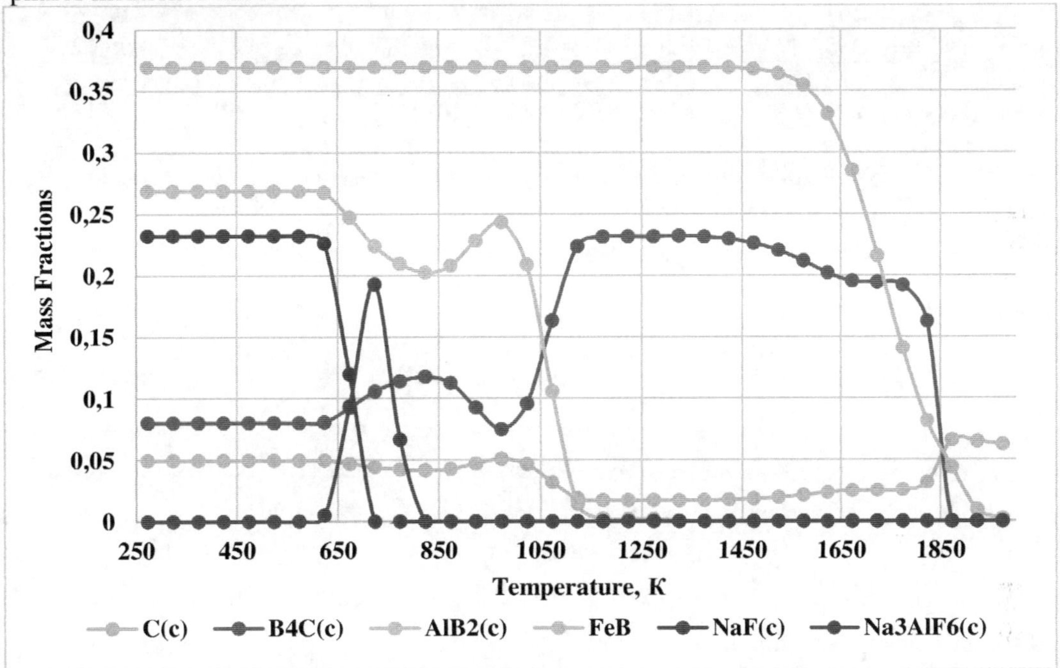

Figure 2. The output of the condensed phases during the synthesis of transition metal borides on low-carbon steel surfaces (T=250-2000 K, P=10^{-3} Pa)

Approximate calculation method. The probability of chemical reactions was determined by the approximate calculation method. The calculation of the chemical reaction equilibrium is estimative, allowing to assess a reaction's potential at a given temperature. Table 3 presents the approximate ΔG values for the following reactions:

$$NaF+B \rightarrow Na+BF \quad (1)$$

$$3NaF+Al \rightarrow 3Na+AlF_3 \quad (2)$$

$$3NaF+AlF_3 \rightarrow Na_3AlF_6 \quad (3)$$

$$2B+Al \rightarrow AlB_2 \quad (4)$$

$$B_4C+2Al \rightarrow 2AlB_2+C \quad (5)$$

$$2Fe+B_4C+2Al+6NaF \rightarrow 2FeB+AlB_2+Na_3AlF_6+3Na+C \quad (6)$$

Table 3. Approximate values ΔG (kJ·mol^{-1})

T, [K]	Reaction 1	Reaction 2	Reaction 3	Reaction 4	Reaction 5	Reaction 6
298.15	416.64	207.9	-87.63	-149.38	-237.43	-2923.51
400	394.34	195.7	-89.09	-148.42	-235.56	-2937.65
600	356.34	184.2	-91.99	-146.08	-231.11	-2950.75
800	320.3	173.74	-96.22	-143.26	-226.72	-2961.83
1000	286.08	179.13	-111.57	-128.65	-199.8	-2952.55
1200	256.04	185.45	-117.02	-123.79	-193.05	-2944.37
1400	261.06	294.74	-142.46	-129.99	-208.65	-2857.04
1600	236.78	314.16	-167.07	-138.17	-227.98	-2866.62
1800	212.77	332.5	-193.14	-151.3	-256.53	-2879.47
2000	188.71	348.94	-220.44	-169.62	-294.45	-2869.15

Approximate calculations was carried out according to the sequence reported in [10]. Standard enthalpy for ΔH°_{298} substances formation and standard S°_{298} entropy were found in reference sources [12-13]. The formulas (Eq. 7) and (Eq. 8) determined the change in enthalpy ΔH°_{T} and entropy ΔS°_{T} as a result of the reaction, and then, using formula (Eq. 9), the change in free energy ΔG_T at temperature T was calculated, as follows:

$$\Delta H^{0}_{\mathrm{T}} = \sum \Delta H^{0\,(\text{products})}_{\mathrm{T}} - \sum \Delta H^{0\,(\text{reagents})}_{\mathrm{T}} \tag{7}$$

$$\Delta S^{0}_{\mathrm{T}} = \sum \Delta S^{0\,(\text{products})}_{\mathrm{T}} - \sum \Delta S^{0\,(\text{reagents})}_{\mathrm{T}} \tag{8}$$

$$\Delta G^{0}_{\mathrm{T}} = \Delta H^{0}_{\mathrm{T}} - T \tag{9}$$

The calculation showed that the last reaction (Eq. 6) is the most probable, considering its high thermodynamic potential.

3. Methodology of Experiment

Low-carbon St3 steel (0.2% C) was used as a test material (analog to A284Gr.D steel). It is used to produce profiled bars and bar irons, sheet and plate steel, cold-rolled, and broadband sheet products. Moreover, this steel is a foundation for pipes (including rectangular pipes), forgings and stampings, tapes, metalware, and wires.

The treatment mixture composition was taken as B_4C -78%, Al-18%, and NaF-4% (wt.) and ethyl alcohol as a binder [10]. The obtained paste was applied to the samples' defatted surface using a brush with a uniform layer of 0,5-1 mm.

Electronic heating with a continuous beam was carried out for 1-3 minutes at a specific power of W=5,7×10^2 W/mm^2 (diameter of the electron beam d=1 mm). The residual pressure in the vacuum chamber did not exceed 2×10^{-3} Pa.

XRD analysis was carried out on a Phaser 2D Bruker diffractometer (Cu Kα - radiation). Microhardness was measured on a PMT-3M microhardness tester. The microhardness tester is equipped with an adapter with a digital camera and the NEXSYS ImageExpert MicroHardness 2 fingerprint image processing program (Russian State Standard 9450-76).

Cross-sections of the samples' microstructure was studied using a METAM RV-21 metallographic microscope equipped with a VEC-335 digital camera and the NEXSYS Image Expert Pro 3.0 software package for quantitative metallographic analysis.

IOP Conf. Series: Materials Science and Engineering 1198 (2021) 012009 doi:10.1088/1757-899X/1198/1/012009

4. Results

The layer with the thickness of 240-260 μm was obtained as a result of EBA (figure 3). The microstructure of the layer consisted of the light zone with embedded dark dendrites. Underneath the layer carbon-rich zone in the form of perlite was obtained due to carbon displacement from the surface by alloying elements. XRD analysis showed the presence of Fe_2B, Fe_3C, $AlFe_3$, and NaF. The last phase is the only one compound that was predicted by calculations. However, it is known that iron borides and aluminides form during the boroaluminizing process or alloying from boron/aluminum-containing powders. The formation of cementite seems to be reasonable due to the boron carbide presence in the treatment paste.

The microhardness measurements showed that the coating hardness was twice as higher as the substrate hardness (figure 3). The maximum value in the coating reached 3800 MPa at around 100 μm from the surface. Just below the coating, the second rise of the hardness was visible in the perlite-rich zone. The relatively low microhardness of the layer was probably due to the low concentration of iron boride.

Based on the experimental results additional chemical reactions were possible in the thermodynamic system:

$$14Fe + B_4C + Al + NaF \rightarrow 4Fe_2B + Fe_3C + AlFe_3 + NaF \quad (10)$$

$$14Fe + B_4C + Al \rightarrow 4Fe_2B + Fe_3C + AlFe_3 \quad (11)$$

$$B_4C + 11Fe \rightarrow 4\,Fe_2B + Fe_3C \quad (12)$$

Table 4. Approximate values ΔG (kJ·mol^{-1})

T, [K]	Reaction 10	Reaction 11	Reaction 12
1600	71.12	-619.2	-362.25

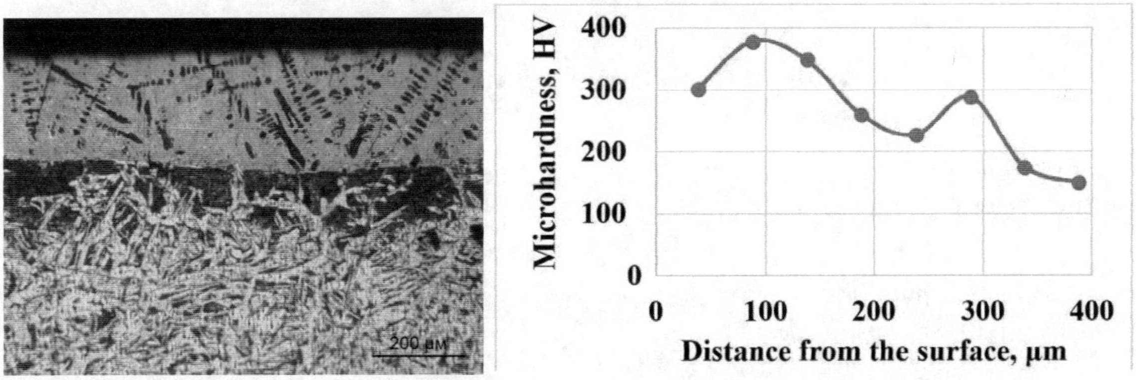

Figure 3. The microstructure and microhardness distribution of St3 steel after EBA

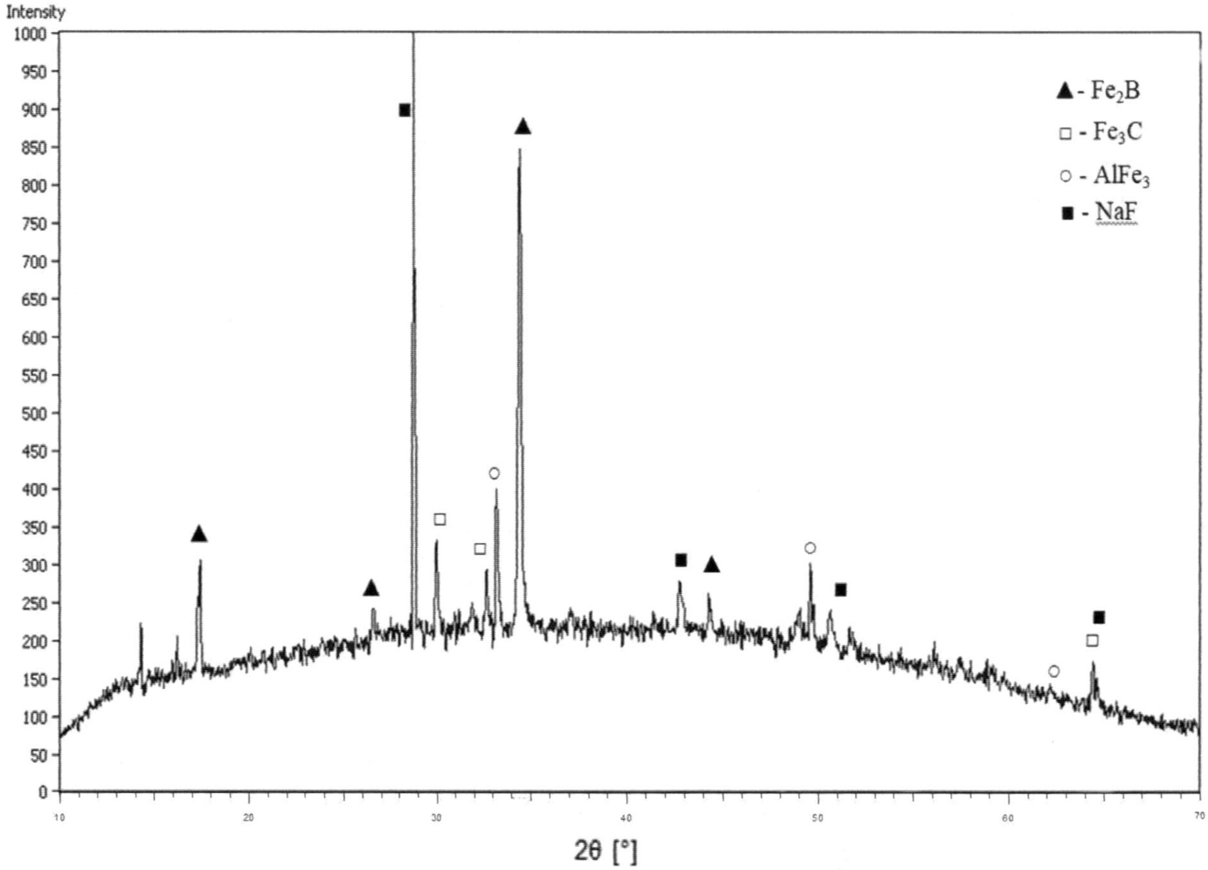

Figure 4. The XRD pattern of St3 steel after EBA

Acknowledgment
The study was carried out with a grant from the Russian Science Foundation (Project No. 19-79-10163).

Conclusion

The presence of activator (NaF) in the treatment paste is redundant for the EBP in a vacuum because XRD analysis has shown its traces after EBA. Besides, the calculation of reactions with sodium fluoride has shown their low probability (positive Gibbs energy), except for the reactions with Na_3AlF_6. However, the latter one wasn't confirmed by XRD analysis.

The interaction of B_4C+Al+NaF reaction mixture with the St3 steel surface in the temperature range from 250 to 2000 K has shown that the predicted coating's phase composition and experimental data differs significantly. Calculation methodology needs to be improved to provide adequate prediction functions. One of the reliable approaches is to utilize verified thermodynamic databases.

References

[1] Węglowski M St, Błacha S and Phillips A 2016 *Vacuum* **130** pp 72–92
[2] Sizov I G, Smirnyagina N N and Semenov A P 2001 *Metal Science and Heat Treatment* **11** pp 45–46
[3] Kulka M 2019 *Trends in Boriding, Engineering Materials, Springer, Cham* pp 99–253
[4] Christiansen T, Bottoli F, Dahl K, Gammeltoft-Hansen N,Laursen M and Somers M 2017 *Materials Performance and Characterization* **6** pp 475–491

[5] Voroshnin L G, Mendeleeva O L and Smetkin V A 2010 *Theory and Technology of Thermal-Chemical Processing* (Minsk: Novoje Znanije) [in Belarus]
[6] Czerwinski F 2012 *Thermochemical treatment of metals, in Heat Treatment: Conventional and Novel Applications* (Rijeka: Czerwinski, Intech Open Access) pp 73–112 [in Croatia]
[7] Mishigdorzhiyn U, Chen Y, Ulakhanov N and Liang H 2020 *Lubricants* **8** p 26
[8] Smirnyagina N N, Sizov I G, Semenov A P et al. 2002 *Metal Science and Heat Treatment* **44** pp 35–38
[9] Tsipas S A, Omar H, Perez F H and Tsipas D N 2008 *Surf. and Coat. Technol.* **202** pp 3263–3271
[10] Mishigdorzhiyn U, Polyansky I, Sizov I, Vetter B, Schlieter A, Heinze S and Leyens C 2017 *Materials Performance and Characterization* **6** pp 531–545
[11] Trusov B G 2002 *Proceedings of the III International Symposium on Theoretical and Applied Plasma Chemistry* [Ples] pp 217–218 [in Russian]
[12] Barin I 1995 Thermochemical Data of Pure Substances **1** Ag-Kr **II** La-Zr By Ihsan Barin (ZEUS GmbH, FRG) in collaboration with Gregor Platzki VCH (New York)
[13] Glushko V P, Medvedev V A, Gurvich L V, Yungman V S, Eds. 1999 The Thermal Constants of Substances (New York: Wiley)

Tunable microstrip liquid crystals patch antennas

N A Romanov[1], T A Chimytov[1,2], A V Nomoev[1*], E Ch Khartaeva[1], E A Gurvitz[3]

[1]Institute of Material Physics of Siberian Branch of Russian Academy of Science, 6 Sakhyanovoi Street, Ulan-Ude, 670047, Russia

[2]Department of Physics and Engineering, Buryat State University, 24a Smolin Street, Ulan-Ude, 670000, Russia

[3]ITMO University, Kronverksky prospect 49, St. Petersburg, 197101 Russia

E-mail: nomoevav@mail.ru

Abstract. A patch antenna based on liquid crystals for the5G GHz range was simulated. The variation of the antenna frequency range after doping magnetic nanoparticles was verified. The high-frequency patch antenna parameters are calculated considering the dependence of the reflectivity of the antenna 11 on the dielectric anisotropy, gain, directivity and phase shift. The ability to control the frequency range of the patch antenna is shown. The phase distributions were demonstrated for 5CB crystals with a slight variation in the dielectric constant tensor and crystals with magnetic permeability.

1. Introduction

The fifth generation of 5G wireless mobile communication provides for the operation of networks at low (up to 2 GHz), medium (2-10 GHz), and high (over 10 GHz) frequencies. Frequencies of 24.5-29.5 GHz are used to transmit signals at a higher speed. In this case, 5G communication is carried out on millimeter waves. Antennas with liquid crystals are used [1,2] along with the known antennas that change their resonance properties due to mechanical changes in size, containing semiconductor devices such as pin-diodes, varactors, ferroelectrics [1]. The advantages of such antennas are a fast and straightforward change of the resonant frequency under the influence of the bias voltage, small dimensions in height, and power consumption compared to semiconductor counterparts. Due to the anisotropy of the dielectric properties of nematic liquid crystals, a change is achieved that is especially pronounced at high frequencies. The use of liquid crystals with increasing frequency leads to a decrease in dielectric losses [4]. The main disadvantages of liquid crystal antennas are low gain and radiation efficiency [3]. Several patches are used to improve these characteristics of antennas, for example, three microstrip sections as emitters and a tunable high-resistance resistor [3]. It is known that the introduction of nanoparticles changes the dielectric anisotropy of liquid crystals [5, 6], which will make it possible to control the position of the resonant frequency of the antenna over a broader range. Therefore, studies of tunable 5G patch antenna parameters based on liquid crystal crystals with embedded nanoparticles are relevant.

2. Modelling Tunable 5G LCD patch antenna

Modeling of a tunable patch antenna operating at a frequency of 28 GHz has been carried out. For this, 5CB liquid crystals were placed in the cavity of the patch antenna. Liquid crystals 5CB have a dielectric loss tangent of 0.03. And also, their complex part of the dielectric constant decreases sharply in the high-frequency region according to the empirical Cole-Cole equation for the case with distributed relaxation time ($\varepsilon\perp\approx2.7$ and $\varepsilon II\approx2.9$) [7,8]. A schematic representation of a patch antenna with designations is shown in Figure 1.1 (a). The exact geometric parameters of the antenna are shown in Table 1.1.

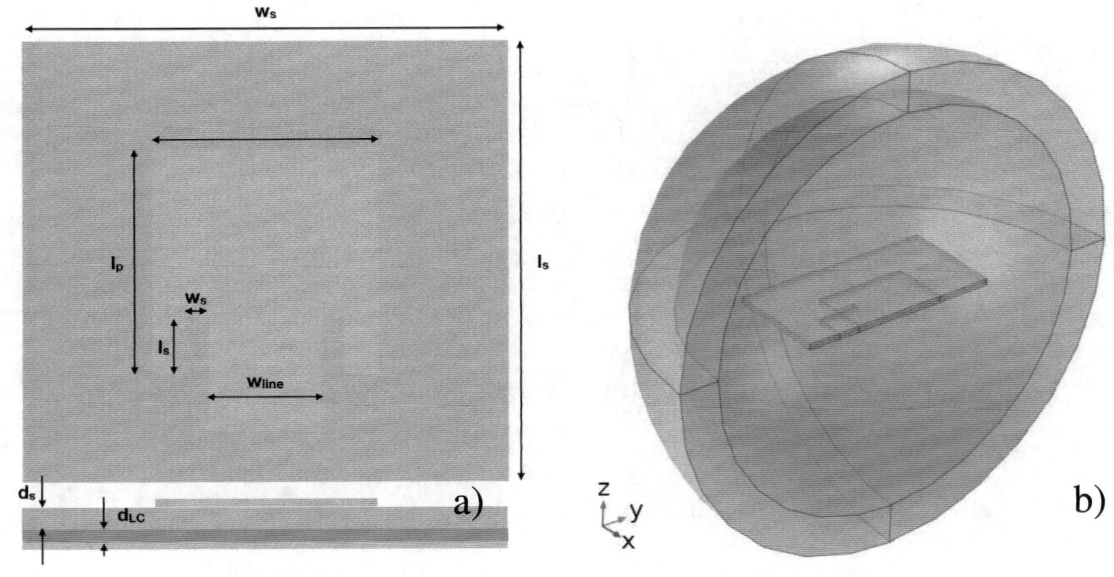

Figure 1.1 - (a) Schematic representation of a patch antenna
(b) Settlement Domain

Table 1.1. Parameters of a tunable 5G patch antenna for phased array

Parameter	Value, mm
Substrate length, l_s	7
Substrate width w_s	7
Patch length, l_p	3.2
Patch width, w_p	4
Construction strip length, l_s	0.7
Building strip width, w_{line}	0.32
Substrate thickness, d_s	0.05
Liquid crystal layer thickness, d_{LC}	0.1
Foil thickness	0.018

The finite element method was used to simulate the patch antenna. The patch antenna is placed in the air domain (Figure 1.1 (b)). The antenna is driven by a discrete port. The copper surfaces in Figure 1.1 (a), marked in yellow, are introduced into the numerical model in the form of the transient boundary condition with a conductivity $\sigma = 56$ MS/m with an adequate thickness of 18 μm. Unlike ideal conductors often used to design antennas, this approach produces more physical results. A boundary condition of the PMC type was used on the plane of the ball section to speed up the calculations due to the mirror symmetry of the antenna geometry.

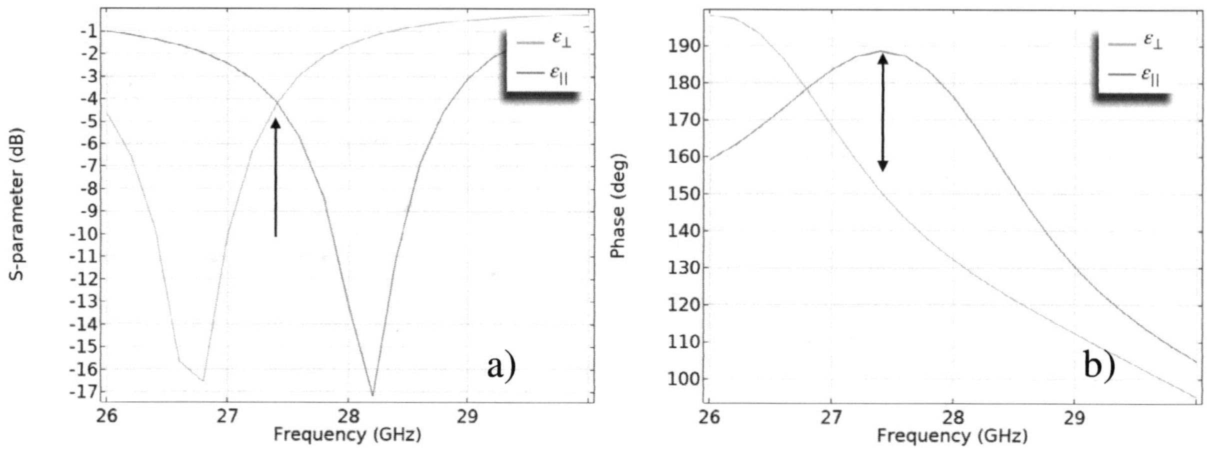

Figure 1.2 (a)Reflection coefficient S11 for two polarizations of LC 5SV
(b)Phase shift of the electric field in the far zone

A rigid thin polyimide film was selected to ensure a high Q-factor of the antenna resonator and a significant phase shift. The real part of the complex dielectric constant of such a film is = 3 (AP9111R or CG352535E), the tangent of the dielectric loss angle tanf_0 (δ) = 0.003, the thickness d_s = 0.1 mm. The patch sizes w_p and l_p are set slightly differently to mismatch the TM mode's excitation by the magnetic field and dominantly excite the TE01 mode by the electric field. The choice of such dimensions is essential to maintain the polarization of the antenna radiation. The parameters l_s and w_s are subscripts and determine the efficiency of radiation input to the antenna. The choice of w_{line} corresponds to the parameters of the microstrip line, which provides radiation input to the antenna to match the source with an impedance of 50 Ohm.

Frequency dependences of the signal reflection coefficient S_{11} were plotted for two spatial crystal orientations. As shown in Figure 1.2 (a), when the orientation of the liquid crystals changes, the resonance shifts, but the directional pattern remains unchanged (Figure 1.3). The phase shift of the electric field in the far zone is about 40 degrees in a reasonably wide frequency band, shown by arrows in Figure 1.2 (a, b).

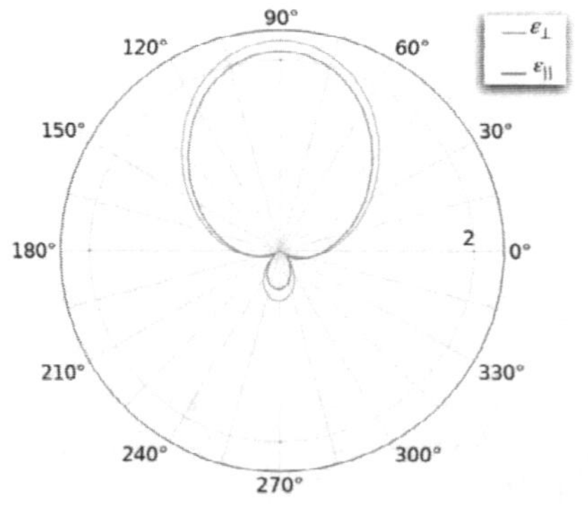

Figure 1.3 – Radiation pattern of the antenna radiation to the far zone.

The maximum efficiency of the received antenna is found not for the center frequency, but along the intersection point of the S11 curves, where the reflectance is at the level of -4 dB. In conclusion, it should be noted that the developed patch antenna is quite effective for creating phased antenna arrays.

IOP Conf. Series: Materials Science and Engineering 1198 (2021) 012010 doi:10.1088/1757-899X/1198/1/012010

The final characteristics of a simple patch antenna are shown in table 1.2.

Table 1.2. Antenna parameters

Parameter	Value
S11	-17dB
Gain	5.7dB
Focus	6.9dB
Maximum far-field phase shift	± 20 degrees

3. Tunable 5G patch antenna based on liquid crystal crystals with the addition of Janus-like magnetic particles

The geometric parameters of the previous antenna were optimized to simulate a tunable patch antenna operating at 28 GHz with Janus-like particles. It is assumed that liquid crystals are controlled by external electric and magnetic fields. The magnetic permeability tensor is added to the numerical model, taking into account the average dispersion of magnetic particles for this range. It is known that the real part of the magnetic permeability of round-shaped iron oxide nanoparticles in the region of 10 GHz and more weakly changes close to 1, and the imaginary part is equal to 0 [9]. The magnetic permeability of composites based on micron-sized powders of carbonyl iron has a pronounced dispersion, the real part in the region of 10-40 GHz varies from 1.7 to 0.7, the imaginary part - from 1.5 to 0.7 [10], and it was found that the dielectric and magnetic permeabilities increase logarithmically with increasing volumetric iron content. Considering that iron oxide nanoparticles can be used in the form of long rods and high range, it can be assumed that the magnetic permeability will change depending on their orientation within the following limits – ($\mu_{\perp} = 1.1$ and $\mu_{II} \approx 1.4$). A schematic representation of a patch antenna with designations is shown in Figure 1.1 (a). Antenna geometric parameters are shown in the table below.

Table 2.1. Parameters of a tunable 5G patch antenna for phased array

Parameter	Value, mm
Substrate length, l_s	7
Substrate width w_s	7
Patch length l_p	3.1
Patch width, w_p	3.9
Construction strip length, l_s	0.65
Building strip width, w_{line}	0.2
Microstrip line width, t	0.48
Substrate thickness, d_s	0.1
Liquid crystal layer thickness, d_{LC}	0.1
Foil thickness	0.018

IOP Conf. Series: Materials Science and Engineering 1198 (2021) 012010 doi:10.1088/1757-899X/1198/1/012010

Similarly to the previous case, to ensure a high Q-factor of the antenna resonator, as well as a higher phase shift, a polyimide film with a dielectric constant $\varepsilon_s = 3$ (AP9111R or CG352535E), a tangent of the dielectric loss angle tan (δ) = 0.003 and a thickness $d_s = 0.1$mm.

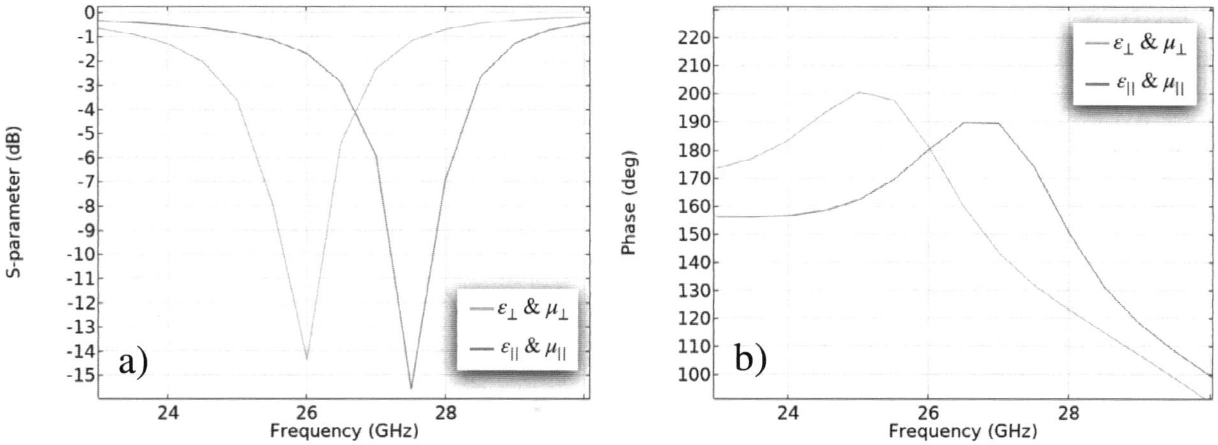

Figure 2.1 - (a) Reflection coefficient S11 for two polarizations of LC 5SB with the addition of Janus-like magnetic particles. (b) Phase shift of the electric field in the far zone.

Frequency dependences of the signal reflection coefficient S_{11} were plotted for two spatial crystal orientations. As can be seen from Figure 2.1 (a), when the orientation of the liquid crystals changes, the resonance shifts, and for the case of LCs with the addition of magnetic nanoparticles, the resonance bands are shifted to the low-frequency region, and the distance between the two S_{11} peaks is reduced by more than 250 MHz.

The phase shift of the electric field in the far zone is of the order of 40-50 degrees, which suggests that the use of magnetic particles to control patch antennas allows achieving a more significant phase shift and, potentially, with an increase in the magnetic permeability anisotropy, to achieve better patch antenna configurations for the 28 GHz range.

As in an antenna without magnetic particles, the maximum phase shift efficiency is found not for the center frequency but for the intersection of the S11 curves, where the reflection coefficient is at a level of ~ -4dB. The final characteristics of a simple patch antenna are shown in table 2.2.

Table 2.2. Antenna parameters

Parameter	Value
S11	-15dB
Gain	5.4dB
Focus	6.7dB
Maximum far-field phase shift	$\pm$25 degrees

4. Conclusion

The results of modeling a patch antenna based on liquid crystals with Janus-like magnetic nanoparticles show that the gain and directivity do not change. The phase shift of the electric field in the far zone exceeds 5 degrees, a similar change for an antenna with undoped liquid crystals. The prospects of using external electric and magnetic fields to control the phase shift of a patch antenna with magnetic nanoparticles are shown.

5. Acknowledgments

This work was supported by the Russian Scientific Foundation [project number 18-79-10143].

References

[1] Kim D, Kim K, Kim H, Choi M and Na J-H, 2021, *Materials*, **14**, 932 https://doi.org/10.3390/ma14040932
[2] Shu J U N; Member G S; Zhang Y, 2020, *IEEE Access*, **8**, 118687–95
[3] Sazegar M, Zheng Y L, Maune H, Damm C, Zhou X ., Binder J, and Jakoby R, 2011 *IEEE Trans. Microwave Theory and Techniques*, V 59 **5**, 1265–73
[4] Deo P, Syahkal D M, Seddon L, Day S E and Fern´andez F A, 2015, *IEEE Trans. Microwave Theory and Techniques*, V 63 **4**, 1388–98
[5] Konshina E A, Gavrish EO, Orlova AO and Artemiev M V, 2011, *JETP Letters* T 37 **21,** 47–54
[6] Ouskova E, Buchnev O, Reshetnyak V, Reznikov Yu, Kresse H, 2003, *Liquid Crystals* V 30 **10**. 1235–39 doi: 10.1080/02678290310001601996
[7] Cole K S and Cole R H, J., 1941, Chem. Phys., V 9 **4**, 341–52
[8] Cole K S and Cole R H, J. 1942, *Chem. Phys.*, V 10 **2,** 98–105
[9] Hong Zh, Yong Y, Fu-Sheng W, Hai-Bo Yi, Dong Zh and Fa-Shen L, 2009, *Chin. Phys. Lett.*, V 26 **1**, 017501
[10] Garanov A, Gibets S G, Politko A A, Semenenko V N and Chistyaev V A, 2018, *Proc. Int. Conf NMMM XXIII*, Moscow, 652-4

IOP Conf. Series: Materials Science and Engineering 1198 (2021) 012011 doi:10.1088/1757-899X/1198/1/012011

Modifying of the dielectric properties of adsorbed water in the force field of a solid surface

I G Simakov, Ch Zh Gulgenov and S B Bazarova*
Laboratory of Molecular Structure Physics, Institute of Physical Materials Science, Siberian Branch of the Russian Academy of Sciences, 670047, Ulan-Ude, Russia

*E-mail: baz_say@mail.ru

Abstract. The results of thermodynamic and acoustoelectric studies of changes in the dielectric properties of adsorbed water under the influence of a solid surface force field are presented. It is noted that the dielectric properties of adsorbed water depend not only on the characteristic size of the system, but also on the interaction energy with the solid surface. Near the hydrophilic solid surface, the density of adsorbed water is increased and the mobility of molecules is reduced. The opposite effect occurs for hydrophobic surfaces. The frequency dependences of the dielectric properties of adsorbed water, as in the case of water in the solid and liquid phase, can be characterized using Debye relaxation theory. It is shown that a significant increase in the orientational dielectric relaxation time in adsorbed water can be explained by taking into account the energy of interfacial boundaries.

1. Introdution

The boundary layer of water adsorbed on the surface of a solid is under the influence of its force field. The molecules of the adsorption layer interact with all the molecules forming the solid phase. However, the direct and most intense interaction is between the molecules of the boundary layer and the molecules forming the surface of the solid. The adsorption boundary layer of a liquid changes the free energy of a solid surface, while the transition layers of a solid can change the charge distribution in the molecules of the boundary layer, causing their structural changes [1, 2].

The change in any structure-sensitive property of a fluid in the boundary state is due to the presence of an orientation induced by hydrogen bonding. As the distance from the solid surface (with increasing layer thickness) the modification of the boundary liquid properties decreases. The thickness of the boundary phase formed near the interface, with a special structure and properties different from those of the bulk liquid phase, varies widely. Both monolayer and multilayer orientations are possible. The structured boundary phase of polar liquids with a multilayer orientation reaches a thickness of the order of several tenths of a micron [3, 4].

The particular structure of the boundary layers of water on the solid surface can be explained by the epitaxial effect of the adsorbent. The active centers of the solid surface impose on adjacent layers of water a special structure (compared to a bulk liquid), which due to the directionality of hydrogen bonds in water extends to a distance of the order of several nanometers from the surface [4].

The adsorption boundary layer is formed on the hydrophilic surface of a solid as a result of polymolecular adsorption of water vapor. The structure of water in the boundary phase differs from its structure in the bulk liquid phase [3]. And as a consequence, the physical properties of adsorbed water (density, viscosity, compressibility, dielectric constant, etc.) differ from the physical properties of water in the bulk phase. A change in the physical properties of adsorbed water is indicated by the fact that the isotherm of polymolecular vapor adsorption $h(p/p_s)_T$ crosses the saturation axis ($p/p_s = 1$) at a finite value of the boundary layer thickness, in contrast to the adsorption isotherms of non-polar liquids [4].

2. Influence of interface energy on the properties of the adsorbed liquid

In order to take into account the change in the interaction energy of the adsorbate molecules as a function of the vapor pressure, hence the distance to the adsorption centers, let us consider polymolecular adsorption under equilibrium conditions. To do this, place a solid with a smooth surface (adsorbent), whose temperature is T_2, in a closed volume (figure1 *a*) above the surface of a liquid with a temperature of T_1. The vapors of this liquid are in thermodynamic equilibrium with the vapor of the liquid in the adsorption layer.

At sufficiently low temperatures, when the vapor above the liquid surface can be regarded as an ideal gas, the liquid – vapor coexistence curve equation is of the form [4]:

$$p_s = p_t \exp(Q_L / RT_1), \qquad (1)$$

where p_s is the saturating vapor pressure, p_t is the thermal pressure of the liquid, $Q_L = N_A E_L$ the latent heat of evaporation (condensation) per mole of matter, N_A is Avogadro's constant. The adsorbed liquid at equilibrium conditions at the same saturating vapor pressure $p_s = p'_s$ has a higher temperature $T_2 > T_1$. This is possible when the latent heat of evaporation of the adsorbed liquid $Q(h)$ is also greater than Q_L , i.e., when $Q(h) = Q_L + \Delta Q$, therefore, we can write that

$$p'_s = p_t \mathrm{e}^{\frac{Q(h)}{RT_2}} \cdot \mathrm{e}^{-\frac{\Delta Q}{RT_2}}. \qquad (2)$$

By equating the right-hand sides of expressions (1), (2) and making the necessary transformations, we obtain

$$\Delta Q = Q_L \frac{T_2 - T_1}{T_1} + RT_2 \ln \frac{p'_t}{p_t}. \qquad (3)$$

In determining the ratio p'_t / p_t, one can use an expression for the thermal pressure, which is obtained using free-volume theory with a rectangular approximation of the cell potential [4]:

$$p_t = \frac{3k_B T}{V_a - V_d}, \qquad (4)$$

where V_a and V_d are the cell volume and the volume occupied by an sample particle.

Under the influence of the solid surface field the density of the adsorbed liquid is increased, we can assume that $V'_a < V_a$, therefore $V_a = V'_a + \Delta V$. Making a number of simplifications, the main of which is that the temperature changes of cell volumes of ordinary and adsorbed liquids are almost identical, we obtain the following expression for the particle adsorption energy $E(h)$ [4]:

$$E(h) = \frac{Q_L}{N_A}\frac{T_2}{T_1} + k_B T_2 \ln\left[\frac{T_2}{T_1}(1 - \alpha_t \Delta T)\right], \qquad (5)$$

where α_t is the coefficient of thermal expansion of the liquid

Equilibrium in energy flow in nanoscale adsorption layers, is established in finite time. The dielectric characteristics tend to equilibrium values during the relaxation process. After the establishment of local equilibrium, the relaxation stage follows, which is described using the hydrodynamic approach. It is assumed that the thermodynamic parameters of the system and the average mass transfer rate do not change significantly during the relaxation time. The adsorption layer, as a thermodynamic system, relaxes to an equilibrium state slowly [4].

Liquid molecules are in continuous thermal motion. Therefore, the dielectric properties of liquids must depend on the dynamic structure of the liquid. The dynamics of molecules is characterized by the relaxation time τ. The relaxation time depends on the intermolecular interaction potential and temperature. Therefore, the relaxation time depends on the dynamic properties of the dipole water molecule and the environmental parameters [4].

Dielectric relaxation in adsorbed water can be interpreted similarly to a phase transition, i.e., a transition from an ordered to an equilibrium state. The thermodynamic Gibbs potential must be complemented by the energy of the interphase boundary. From the condition of equality of the ordered and equilibrium state potentials we obtain

$$\delta T = 1 - \frac{T}{T_0} = \frac{\Delta\sigma_s + \Delta\sigma_f}{\Delta F \cdot h} \tag{6}$$

where T_0 and T are temperatures of the liquid in the bulk and boundary phases, corresponding to the same dielectric relaxation time ($\tau_1 = \tau_2$); $\Delta\sigma_s$ and $\Delta\sigma_f$ are changes in the specific surface energy of the "solid-liquid" and "liquid-vapor" boundaries during relaxation; ΔF is the relaxation process activation energy, h is the adsorption layer thickness.

The dielectric relaxation time is related to the change in the free activation energy of dipole relaxation ΔF by the following relation [4]:

$$\tau = \frac{2\pi\hbar N_A}{RT} \exp\left[\frac{\Delta F}{RT}\right] \tag{7}$$

where R – gas constant; $\hbar$ – reduced Planck constant.

Substituting the activation energy of the relaxation process obtained from expression (7) into equation (6), we can write

$$\frac{T}{T_0} = 1 - \frac{\Delta\sigma_s + \Delta\sigma_f}{h \cdot RT \cdot \ln\left(\frac{\tau RT}{2\pi\hbar N_A}\right)} \tag{8}$$

Thus, the energy of the liquid-solid interface σ_s in equation (8) characterizes the effect of the force field of the solid surface. In addition, the force field of the solid surface affects the formation of the adsorption layer. In equation (8), this influence is expressed through the thickness of the adsorption layer h.

3. Results and Discussions

Experimental results of measurements of dielectric parameters of adsorbed water as a function of relative vapor pressure at fixed values of SAW frequencies are shown in Figure 1. Curves *1, 3* correspond to frequency f = 62 МГц, and curves *2, 4* – 440 MHz. The character of the dependence indicates the presence of a relaxation process with a relaxation time $\tau \sim 10^{-8}$ s. The closest in time relaxation process (with a relaxation time of the order of 10^{-10} s) in bulk water is the Debye dielectric relaxation [4]. Dielectric relaxation at fixed values of SAW frequencies can be explained by structural-kinetic changes in adsorbed water. As the thickness of the adsorption layer increases, the real part of the complex permittivity changes from the value corresponding to the high-frequency relaxation limit ε_∞ to the value of the static permittivity ε_s of bulk water.

As the frequency of SAW increases, the region of relaxation change in the dielectric permittivity of adsorbed water shifts to the region of higher values of vapor pressure. The character of the curve remains almost unchanged. Since the dependence of the adsorption layer thickness on the vapor pressure p/p_s is gradual, therefore, the same change in dielectric permittivity at higher frequency corresponds to a larger area of change in the thickness of the adsorption layer.

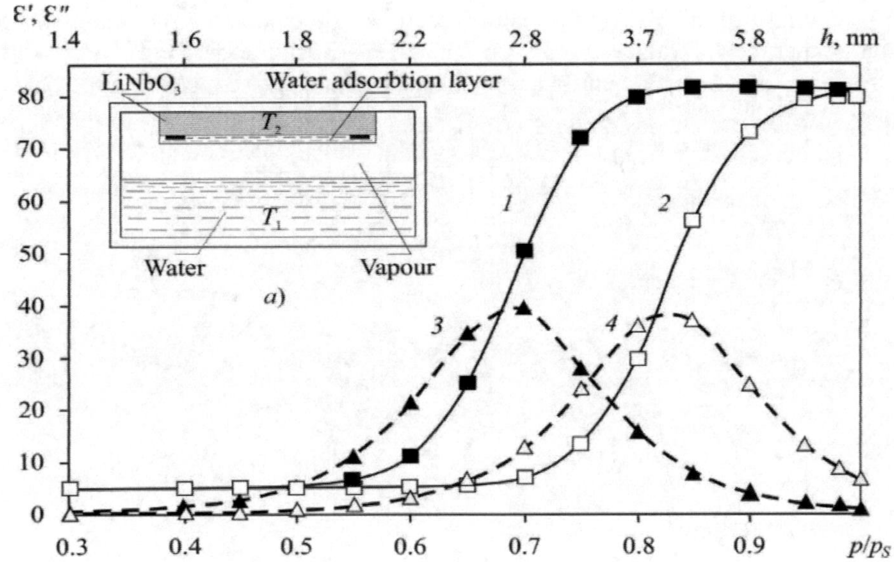

Figure 1. Dependence of the real ε′ (solid line) and imaginary ε″ (dashed line) parts of the complex dielectric permittivity of adsorbed water on vapor pressure and thickness of the adsorption layer (*1*, *3* – 62 MHz; *2*, *4* – 440 MHz).

An experimental study of the dielectric permittivity dispersion of boundary water at fixed values of the adsorption layer thickness showed that the Debye relaxation frequency is significantly lower than that of bulk water (Figure 2, curve *4*).

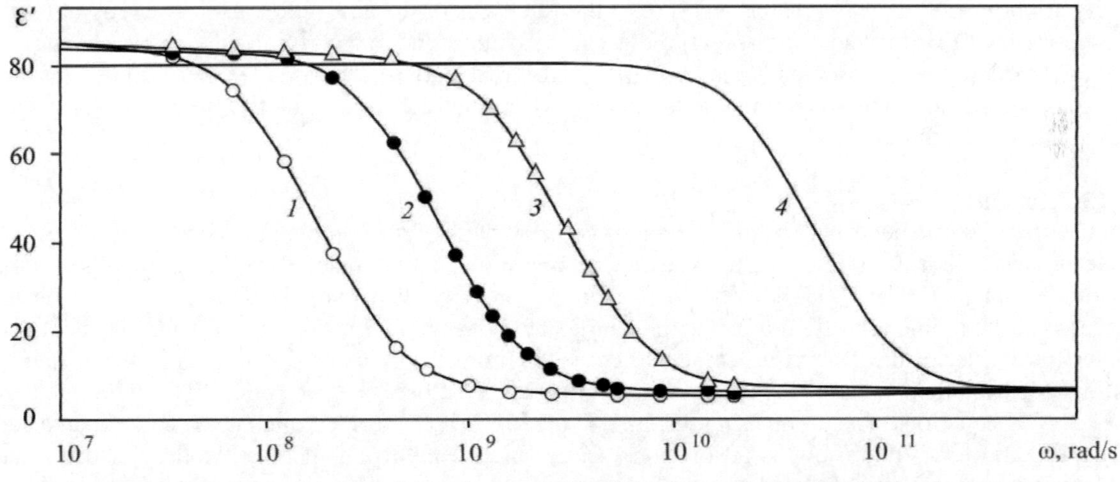

Figure 2. Dispersion of the real part of the dielectric permittivity of adsorbed water.

Figure 2 shows the dispersion of dielectric permittivity of adsorbed (at values of relative vapor pressure: *1* – ($p/p_s = 0.6$); *2* – ($p/p_s = 0.7$); *3* – ($p/p_s = 0.8$)) and bulk water (*4*). Numerical estimation was performed in the frequency range 20 – 400 MHz using the real part of the Debye equation. The markers indicate experimental data on the dielectric permittivity of adsorbed water calculated using the equation [4, 5]

$$\varepsilon' = \frac{2(\varepsilon_p + 1)^2}{K^2 \varepsilon_p} \left(\frac{\Delta V}{h\omega} - \frac{\rho_f}{2\pi\rho_s} \sqrt{1 - V_f^2 / V_s^2} \right). \quad (9)$$

The frequency dependence of the imaginary part of the complex permittivity of adsorbed water is shown in Figure 3. Curves *1*, *2*, *3* of dispersion of dielectric permittivity of adsorbed water at values of relative vapor pressure, respectively, 0.6; 0.7 and 0.8 are constructed using the imaginary part of the Debye equation. Experimental values (points on curves 1 – 3, figure 3) were determined from the experimental isotherms of attenuation and relative change in the rate of SAW using the equation [4, 5].

$$\varepsilon'' = \frac{(\varepsilon_p+1)^2}{K^2\varepsilon_p}\left(\frac{2\alpha V_s^2}{h\omega^2} - \frac{\rho_f V_f \beta}{\pi\rho_s\omega\sqrt{1-V_f^2/V_s^2}}\right). \tag{10}$$

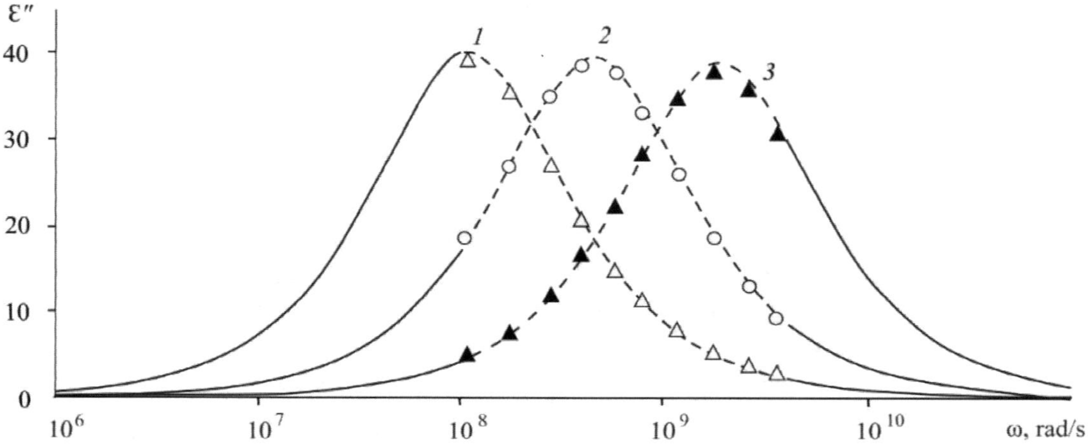

Figure 3. Dispersion of the imaginary part of the complex permittivity of adsorbed water.

It follows from the analysis of the figures that the dispersion of the dielectric permittivity of water in the adsorbed layer is well enough described by Debye's theory. Thus, it has been demonstrated that with variations in the relative vapor pressure or the thickness of the adsorption layer at fixed values of the frequency of surface acoustic waves, dielectric relaxation due to structural and kinetic changes in the adsorbed water is observed. It is shown that the real part of the complex permittivity, as the thickness of the adsorption layer increases, changes from the high-frequency to low-frequency values of the Debye relaxation.

4. Conclusions

The frequency dependences of the real and imaginary parts of the complex dielectric permittivity of water in adsorption layers of fixed thickness were investigated experimentally. It is shown that in the frequency range (20 – 400 MHz) the dielectric permittivity dispersion is detected. A significant increase in the dipole relaxation time can be caused by a change in the structure of water in the layer, under the influence of the crystal surface force field. From the analysis of the experimental data it follows that the dispersion of the dielectric permittivity of adsorbed water is described quite well by the Debye equation. The structuring effect of the adsorbing surface increases with decreasing thickness and temperature of the adsorption layer, which leads to an increase in the viscosity and dielectric relaxation time of the adsorbed water.

References

[1] Churaev N V, Sobolev V D 2011 *Colloid Stability: The Role of Surface Forces* **2** 127 – 152
[2] Rusanov A I 2020 *Colloid Journal* **82** 62 – 8
[3] Churaev N V 1996 *Colloid Journal* **58** 681 – 693
[4] Simakov I G, Gulgenov Ch Zh 2019 *Akustoelektricheskie metody issledovaniya zhidkostej v granichnoj faze* (Novosibirsk: Izd-vo SO RAN)
[5] Simakov I G, Gulgenov Ch Zh, Bazarova S B 2019 *IOP Conf. Series: Journal of Physics* **1281** 012073

IOP Conf. Series: Materials Science and Engineering 1198 (2021) 012012 doi:10.1088/1757-899X/1198/1/012012

Determination of the complex dielectric permeability of liquid at the radiation frequency 32 GHz

B G Tsydypov[1], A V Bazarov[1] , B V Basanov[1], B Z Garmaev[1], E Yu Korovin[2] and E B Atutov[1]

[1] Laboratory of radiosounding of natural media, Institute of Physical Materials Science of the Siberian Branch of the Russian Academy of Science), Sakhyanova street 6, Ulan-Ude, Russian Federation
[2] Radiophysics faculty, National Research Tomsk State University, Lenina street 36, Tomsk, Russian Federation

Abstract. Method and device for measuring the complex dielectric constant of liquids are described. The real part of the complex dielectric constant calculated from the conditions for the maximum and minimum of the measured interference dependence of the reflection coefficient on the layer depth. The imaginary part determined by selection to the maximum coincidence of the calculated and measured dependences of the reflection coefficient.

1. Introduction
The complex dielectric constant of liquids in the millimeter region of radiowave, as a rule, waveguide methods are used [1-5], in which the measured parameters are the phase and level of the reflected or transmitted signal.
Previously proposed method for determining complex permittivity liquid dielectrics, based on knowledge of only the character of the behaviour of the reflected field, depending on the width of the layer of the investigated liquid [6]. The essence of the invention consisted in comparison of the theoretical dependence of the reflection coefficient on the width of the probed layer with the measured one. In the calculation, we used the formula for three layers, where the first layer is air, the second layer is the liquid under study, and the third is the metal layer [7].
Subsequently, the comparison of the reflection coefficient excluded, and instead a formula for determining the dielectric constant proposed. The proposed method tested in the Ka- band for liquids without losses and registered [8]. In this paper, we will show the applicability of the developed method for determining the complex dielectric constant of semiconducting liquids.

2. Determination of complex permittivity
The scheme and photo of the experimental setup shown in Figure 1.

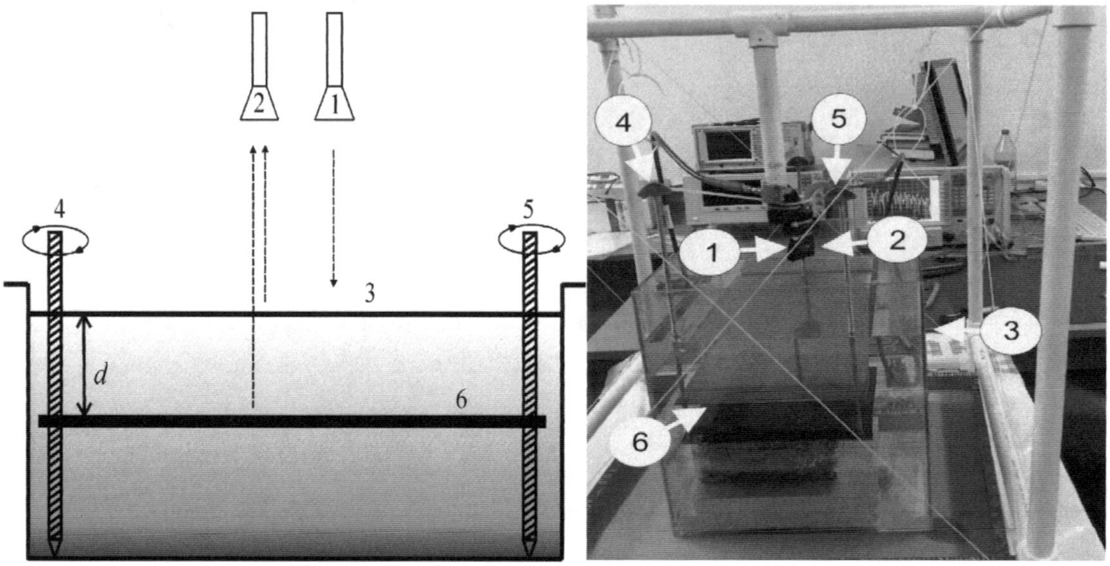

a) scheme b) photo

Figure 1. Measurement setup

Horn antennas emit (1) and receive (2) a linearly polarized electromagnetic wave along the normal to the liquid surface (3), and by turning the guide pins (4 and 5), the position of the metal plate (6) provided - the thickness of the probed layer. Turning the guides through an angle of 45° with a thread pitch of 1.5 mm will correspond to raising or lowering the insert by 1.5 mm / 8 = 0.1875 mm.

In Figure 2 shown the measured dependence of the reflected signal on the immersion depth of the plate at a radiation frequency of 32 GHz. Vegetable oil was chosen as the test liquid.

The real part is determined from the interference condition. For $2k_2 d_{\max} = 2m\pi$ maximum intensity will be observed. Here $k_2 = k_1\sqrt{\varepsilon_2' - i\varepsilon_2''}$ is wavenumber of layer, $d_{\max}$ is layer thickness at which the reflected signal with maximally intensity, m - positive integer. And for $2k_2 d_{\min} = (2m+1)\pi$ we can see minimum ($d_{\min}$ is subsequent after $d_{\max}$ layer thickness at which a minimum is observed). Assuming that $\varepsilon_2' >> \varepsilon_2''$, we compose a system of equations and excluding m and we get

$$\varepsilon_2' \approx \left(-\frac{c}{4f(d_{\max} - d_{\min})}\right)^2 . \tag{1}$$

Here f is frequency of irradiating field, c is velocity of light in vacuum.

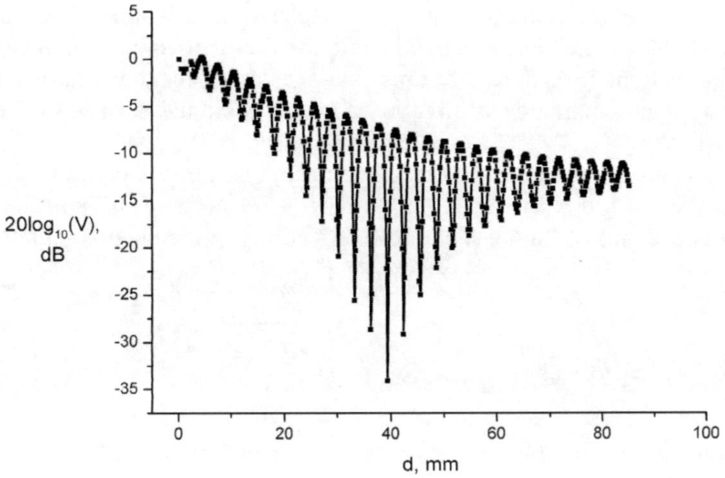

Figure 2. Experimental dependence of the reflected wave intensity on the immersion depth of the plate

The table shows the values d_{max} and d_{min} from Figure 2 are calculated according to (1) values of the real part of complex permittivity.

Table 1. Values of the layer width at which the maxima and minima of the reflection coefficient observed.

d_{max} (mm)	d_{min} (mm)	ε_2'
10.5	12	2.44
13.5	15	2.44
16.6875	18.1875	2.44
19.6875	21.1875	2.44
22.6875	24.1875	2.44
25.6875	27.1875	2.44
28.6875	30.1875	2.44
31.6875	33.1875	2.44
34.875	36.1875	3.18
37.875	39.375	2.44
40.875	42.375	2.44
44.0625	45.5625	2.44
47.0625	48.5625	2.44
50.0625	51.5625	2.44
53.25	54.75	2.44
56.25	57.75	2.44
59.25	60.9375	1.92
62.4375	63.9375	2.44
65.4375	67.125	1.92
68.625	70.125	2.44
71.625	73.3125	1.92
74.8125	76.5	1.92
81	82.3125	3.18

Accordingly, the average value is defined as 2.413913 with standard deviation 0.313423024.

IOP Conf. Series: Materials Science and Engineering 1198 (2021) 012012 doi:10.1088/1757-899X/1198/1/012012

The behaviour of the curve in Figure 2 is the result of the interference of waves reflected from two boundaries: the surface of the liquid and the plate. As the plate deepens, these waves begin to be comparable in amplitudes, and we can observe an increase in the oscillations of the reflection coefficient. Then, the wave reflected from the plate attenuated, and the wave reflected from the surface of the liquid contributes to the reflection. It is quite obvious that the larger the imaginary part of the complex dielectric constant of the medium under study, the closer to the liquid surface the deepest minimum of these fluctuations will be.

We will use this property to determine the imaginary part of complex permittivity. Let us determine imaginary part from the selection to the maximum coincidence of the position of the deepest minimum of the interference dependence of Figure 2 with the reflectivity of the layer calculated by the formula

$$V^2 = \left(\frac{V_{12} + V_{23}\exp(-2ik_2 d)}{1 + V_{12}V_{23}\exp(-2ik_2 d)} \right)^2 . \qquad (2)$$

Where $V_{12} = \dfrac{k_1 - k_2}{k_1 + k_2}$, $V_{23} = \dfrac{k_2 - k_3}{k_2 + k_3}$ are reflection coefficients, $k_1 = \omega / c$, $k_2 = k_1\sqrt{\varepsilon_2}$, $k_3 = k_1\sqrt{\varepsilon_3}$ are wavenumber in media, $\omega = 2\pi f$, $\varepsilon_2 = \varepsilon_2' - i\varepsilon_2''$ is complex dielectric constant of the test liquid, $\varepsilon_3 = \varepsilon_3' - i\varepsilon_3''$ complex dielectric constant of the metal plate, d is the thickness of the probed layer, i - imaginary unit [7].

Figure 3 shows the calculation according to (2) for an irradiation frequency of 32 GHz and $\varepsilon_2 = 2.41 - i0.089$.

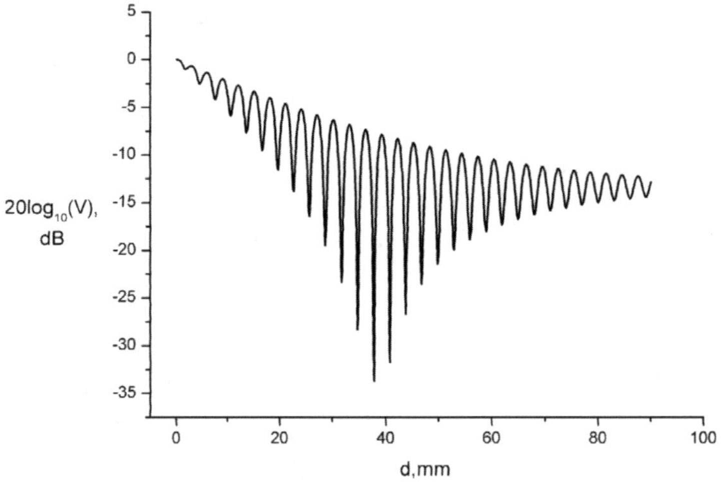

Figure 3. Calculated dependence of the reflected wave intensity on the immersion depth of the plate

Complex dielectric constant of the test oil was also measured by the open waveguide method using an E8363B radio spectroscope from Agilent Technologies, which showed that $\varepsilon_2 = 2.43 - i0.16$. Hence, the relative difference obtained by the two methods is 0.8% and 44.3%, respectively, for the real and imaginary parts.

On Figure 4 the calculation for $\varepsilon_2 = 2.43 - i0.16$ and the frequency of the irradiating field is 32 GHz are presented.

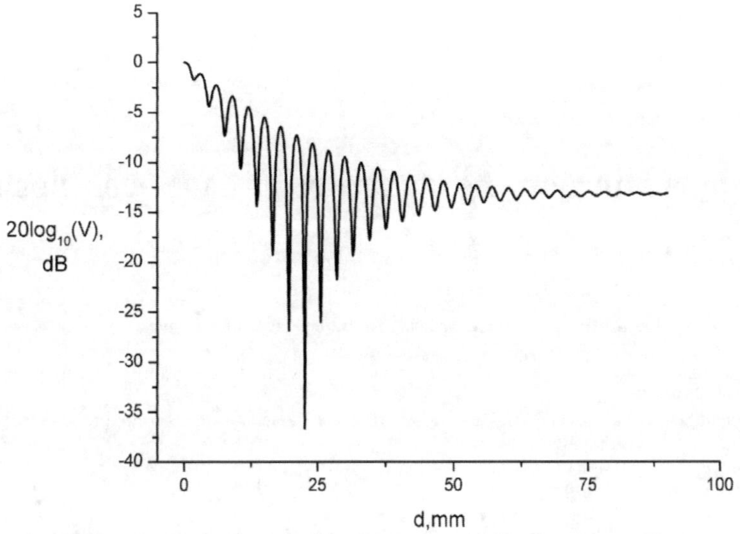

Figure 4. Calculated dependence of the reflected wave intensity on the immersion depth of the plate

A significant difference from the experimental data presented in Figure 2 has been seen.
In [7], it was mentioned that it is possible to experimentally determine the velocity and absorption of acoustic waves from the measured function of reflectivity on the layer width. Analogy of acoustics with radiophysics in the case of radio waves, the authors have not encountered such an approach.
Advantage of the proposed method is that he is not expensive because a scalar field analyser is sufficient for measurements. In addition, the method does not require precision calibration procedures.

3. References

[1] Sobenin N P, Milovanov O S 2007 *The technique of superhigh frequencies. Textbook for universities.* Moscow. Energoavtomizdat Publication. p. 510 (In Russian)

[2] Gusev Yu A 2008 *Fundamentals of dielectric spectroscopy. [Textbook of the Physical Faculty of Kazan State University].* (In Russian).

[3] Dmitriev M S, Dyakonov M V, Guchkin A S, Krasnokutsky R A 2018 *Installation for measuring the complex dielectric constant of liquid dielectrics // Instruments and experimental technique. No.3.* pp. 51-53. DOI: 10.7868/S0032816218030229.

[4] Vaughan W E, Smyth C P 1961 *J. Phys. Chem., № 65,* p. 98.

[5] 01.07.2020 *Agilent 85070E Dielectric Probe Kit [online]. Keysight tehnologies.* Available at: https://www.keysight.com/en/pd-304506-pn-85070E/dielectric-probe-kit?cc=RU&lc=rus.

[6] Patent of Russian Federation 2688825. Atutov E B, Lomukhin Yu L, Basanov B V *Sposob beskontaktnogo izmereniya kompleksnoy dielektricheskoy pronitsayemosti poluprovodyashchikh zhidkostey [Method for non-contact measurement of complex dielectric constant of semiconducting liquids].* The applicant and patent holder: Federal State Budgetary Institution of Science, Institute of Scientific Materials, Siberian Branch of the Russian Academy of Sciences (IPMS SB RAS). Applied 11.04.2018. Published 22.05.2019. (In Russian).

[7] Brekhovskikh L M *Waves in layered media.* Moscow. Science Publication. 1973. p 344 (In Russian).

[8] Patent 2728250. Atutov E B, Evtifeev A A, Ochirov O N, Bazarov A V *Sposob beskontaktnogo opredeleniya dielectricheskoy pronictsaemosti jhidkih dielectrikov v Ka diapazone [Method for contactless determination of dielectric constant of liquid dielectrics in the Ka- diapason].* The applicant and patent holder: Federal State Budgetary Institution of Science, Institute of Scientific Materials, Siberian Branch of the Russian Academy of Sciences (IPMS SB RAS). Applied 20.02.2020. Published 28.07.2020. (In Russian).

IOP Conf. Series: Materials Science and Engineering 1198 (2021) 012013 doi:10.1088/1757-899X/1198/1/012013

Effect of temperature on the structure of Cu/Si composite Janus nanoparticles

N V Yumozhapova[1,2], A V Nomoev[1,2*], N A Romanov[1], E Ch Khartaeva[1,2]

[1]Institute of Material Physics of Siberian Branch of Russian Academy of Science, 6 Sakhyanovoi Street, Ulan-Ude, 670047, Russia
[2]Department of Physics and Engineering, Buryat State University, 24a Smolin Street, Ulan-Ude, 670000, Russia

E-mail: nomoevav@mail.ru

Abstract. This work determined the effect of temperature on the formation of composite Cu/Si nanoparticles by the molecular dynamics (MD) method. For the first time, the temperature dependence of the potential energy of hybrid nanoparticles, which were single-crystal spherical clusters cut from an ideal lattice, was studied. It is shown that phase transitions and structural ordering cause a sharp change in the potential energy of composite nanoparticles. The structural arrangement of copper nanoparticles at high temperatures was found.

1. Introduction

Metal-semiconductor composite nanoparticles with a Jinus-like structure are of particular interest. Firstly, the Janus-like metal/semiconductor nanoparticles created by the gas-phase method are already electronic devices-nanoscale diodes due to the contact between its components formed during the creation of the particle. Secondly, nanoscale diodes, for example, in the form of nanopowders or suspension, can have a high concentration per unit volume, promising for creating new microelectronic devices with a record high density of components. The electrophysical properties, interaction with high-frequency electromagnetic radiation of a Janus-like metal/semiconductor nanoparticle will be strongly influenced by its composition, interface, structure, shape and size, and the dielectric properties of its semiconductor part and environment [1]. When synthesizing nanoparticles, it is necessary to take into account that the formation of a set of composite nanoparticles occurs under conditions of self-organization of the substance, which may result in a greater or lesser deviation of the composition and morphology of individual nanoparticles from the certain averaged (set) values [2, 3]. To solve this problem, we can use the method of heat treatment of the resulting nanoparticles as the final synthesis stage from the gas phase. In addition, one of the known methods for producing metal nanoparticles with an oxide shell is their annealing at elevated temperatures in an oxygen-containing medium. Understanding the structure of the core created by such heat treatment is an important task.

One of the first works on studying the sintering process [4] was published in the 40s of the 20th century. In this and subsequent works, mainly macroscopic particles and, to a lesser extent, microscopic particles were studied [4-9]. The sintering mechanism of solid nanoparticles has not been fully understood despite long-term scientific research in this area. The self-preserving sintering model in [9] suggests that minimization of the agglomerate surface is the main driving force of the sintering process.
Based on these studies, models of the interaction between particle fusion and agglomerate coagulation have been developed [10, 11].

In [12], for the first time, Cu nanoparticles synthesized from the gas phase and, consequently, having different sizes, shapes, and degrees of agglomeration were used to study the processes of thermal exposure.
The morphology of bimetallic nanoparticles can change due to their heat treatment [13-16]. Depending on the composition and structure of the initial nanoparticles, the temperature and the duration of heating of the sample, both segregation of the components and their more uniform distribution over the particle volume is possible. It was shown that the heat treatment of electrocatalysts at 350°C leads to the destruction of the core-shell architecture and the fusion of some bimetallic nanoparticles with the formation of both

homogeneous and ordered solid solutions of PtCu, as well as to the partial reduction of copper oxides (Figure 1) [17].

In [18], the parameters of thermal stability of Cu/Si core-shell composite nanoparticles of different sizes and structures were determined using molecular dynamics calculations.

However, in all the works presented above, nanoparticles were studied, monocrystalline spherical clusters were cut out of an ideal lattice. In an actual experiment, the creation of nanoparticles usually occurs under nonequilibrium conditions, which leads to the formation of various structural imperfections of the particle. This leads to twinning, agglomeration of point defects, and the appearance of lattice types that are not characteristic of matter in macroscopic states.

In our work, when studying the processes of thermal exposure, for the first time, Janus-like Cu/Si nanoparticles synthesized from the liquid phase will be used.

Accurate information on the morphological changes of the investigated particles during heating is necessary to create the most efficient heat-treatment process. For this purpose, two methods of studying the heat treatment of particles are experimentally used - differential electric mobility analyzers (DMA) and transmission electron microscopy (TEM) [19]. These methods are suitable for determining particles' concentration, diameter, and morphology before and after heat treatment. Still, they cannot provide complete information about the processes that occur during the heating of materials and the phenomena that affect the structure and shape of the processed particles. Only computer simulations can provide such information in full. Therefore, the article presents the results of modeling the heat treatment of Janus-like Cu/Si nanoparticles to study the effect of temperature on the production of particles with different, including defect-free shapes and internal structures.

2. Results of Simulation and Discussion

One of the main features of the synthesis is the significant heterogeneity of the internal structure and external shape of the resulting nanoparticles. Therefore, depending on the parameters of the evaporation-condensation process, up to 90% of particles with a pronounced non-spherical shape and different internal structures can be formed in the system. However, the polydispersity of the resulting particles does not always negatively affect the further technological application of this material. Recently, such free particles have been effectively used in various chemical reactions due to their high catalytic activity.

Creating the most efficient heat treatment process requires accurate information on the morphological changes of the particles under study during heating. Modeling of thermal effects on Cu/Si particles of various sizes was carried out by the molecular dynamics method in the LAMMPS software package (Large-scale Atomic / Molecular Massively Parallel Simulator). The modified many-particle potential of the embedded atom MEAM (Modified Embedded Atom Method) was used in these calculations [20]. The number of atoms in the cluster is 1800.

$$U_i = \sum_i F_i(\rho_{h,i}) + \frac{1}{2}\sum_i \sum_{j\neq i} u_{ij}(r_{ij}) \tag{1}$$

where $\rho_{h,i}$ is the electron density at the node of atom i; F_i is the energy of "embedded" of atom i in the electron density of the medium; u_{ij} is the central potential of the pair interaction between atoms i and j located at a distance of r_{ij} from each other.

The "embedded" function:

$$F(\rho_i) = AE_c \frac{\rho_i}{\rho_i^0} ln\left(\frac{\rho_i}{\rho_i^0}\right) \tag{2}$$

where A is the fitting parameter, E_c is the cohesion energy, and ρ_i^0 is the background electron density for the node of atom i in its reference (usually equilibrium) lattice.

Figure 1 shows a graph of the temperature dependence of the potential energy of a janus-like Cu/Si particle during heating in the temperature range of T = 300 K to T = 1600 K during the time t = 10 ns. Figure 2 also shows the evolution of the particle structure under heating as a function of temperature. Moreover, I is the view of the particle from the side from which the copper (black) and silicon (gray) parts of the nanoparticle are visible; II, III, IV are the view of the particle from the copper side at T = 300 K, 850 K, 1400 K, respectively. From the graph and images of the particles, it is seen that no significant changes occur in the amorphous particle in the temperature range from 100 K to 600 K. At T = 650 K, a decrease in energy is observed due to the course of an exothermic reaction and caused by the ordering of copper with the formation of crystal structures. This change is not observed when a copper nanoparticle (created from the gas phase) melts in the case of a nondefective lattice [12]. When the temperature reaches 850 K (particle III), there is a sharp increase in energy due to the melting process of copper. The silicon atoms begin to envelop the copper. When the program runs repeatedly, the copper comes to the surface, forming a dendritic structure.

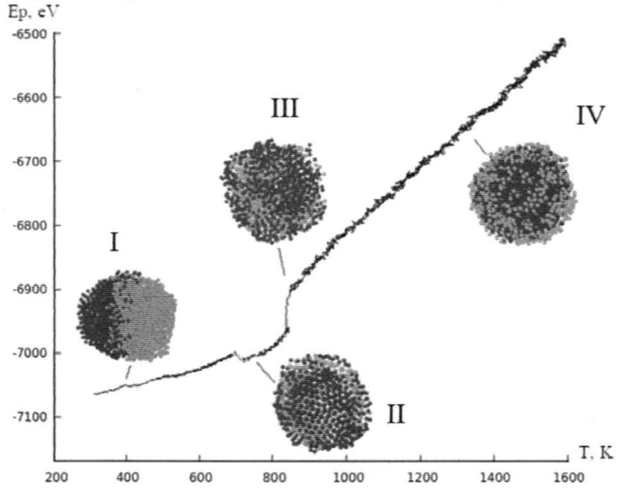

Figure 1. Graph of the temperature dependence of the potential energy of a janus-like Cu/Si particle during heating in the temperature range of T = 300 K to T = 1600 K during the time t = 20 ns. I - view of the particle with the visible the copper (black) and silicon (gray) parts of the nanoparticle are visible, II, III, IV- particle at T = 300 K, 850 K, 1400 K, respectively.

In the case of very slow heating up to 800K for 20 ns the copper formed a defective icosahedral structure. Appearance of a Janus-like nanoparticle Cu/Si to and after thermal exposure is shown in Fig.2.

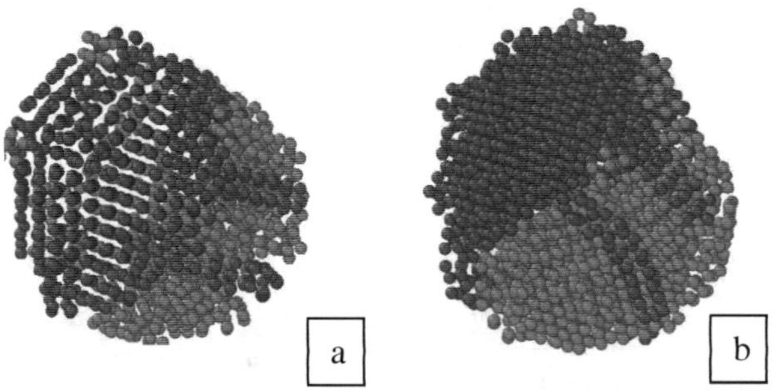

Figure 2. Appearance of the janus-like Cu/Si nanoparticle before - (a) and after thermal exposure - (b)

The described evolution of the structure with temperature changes is confirmed by the radial distribution function of the Cu/Si nanocluster atoms. Figure 3 shows graphs of the radial distribution function of the atoms of a janus-like particle nanocluster at two temperatures. At a temperature of T= 300 K, copper and silicon are in an amorphous state after cooling from the liquid phase (Figure 3a). In Figure 3b, for copper, temperature T= 800 K, the appearance of the second and third peaks is noticeable, while silicon continues to be in an amorphous state.

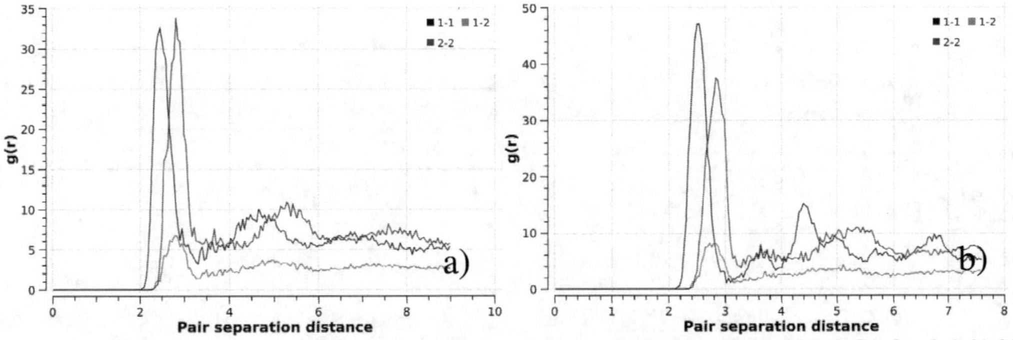

Figure 3. Function of the radial distribution of atoms of Cu/Si nanoclusters (1-1 Cu-Cu; 2-2 Si-Si: 1-2 Cu-Si): a) T = 300K; b) T = 800K.

3. Conclusion

The temperature dependence of the structure of Janus-like metal/dielectric Cu/Si nanoparticles was determined by the numerical method of molecular dynamics using the potential of an embedded atom. The temperature dependence of the potential energy of composite nanoparticles, which are single-crystal spherical clusters cut from an ideal lattice, is investigated. At T = 650 K, an exothermic process occurs, caused by the ordering of copper with the formation of crystal structures. When the temperature reaches 850 K, an endothermic process occurs due to the melting of copper.

4. Acknowledgments

This work was supported by the Russian Scientific Foundation [project number 18-79-10143].

References

[1] Pryadchenko V V, Srabionyan V V,. Mikheykina E B, Avakyan L A , Murzin V Y, Zubavichus Y V, Zizak I, Guterman V E and Bugaev L A., 2015, *J. Phys. Chem.,* **3217**, 119.

[2] Guterman V E, Belenov S V, Pakharev A Y, Min M, Tabachkova N Y, Mikheykina E B, Vysochina L L and Lastovina T A, 2016, *Int. J. Hydrogen Energy*, **41**, 1609

[3] Sauerwald F, 1943, *Kolloid-Zeitschrift*,. V 104 **2–3,** 144–160.

[4] Frenkel Ya, 1946, *JETP,* V 16 **1,** 29–38.

[5] Geguzin Ya E, 1984, *Sintering physics*, (Moscow: Sience)

[6] Pines B Ya, 1954, *Advances in physical sciences,* V 52 **4,** 501–559.

[7] Ivensen V.A. *Sintering phenomenology and some theoretical questions*, 1985, (Moscow: Metallurgy)

[8] Koch W and Friedlander S K, 1990, *Journal of Colloid and Interface Science,* V 140 **2,** 419-427.

[9] Kruis F E et al., 1993, *Aerosol science and technology*, V 19 **4**, 514-526.

[10] Seto T, Shimada M and Okuyama K, 1995, *Aerosol Science and Technology,* V 23 **2**, 183-200.

[11] Chepkasov I V, Gafner Y Y and Gafner S L, 2016, *Journal of Aerosol Science*, **91**, 33-42.

[12] Bezerra C W B, Zhang L, Liu H, Lee K, Marques A L B, Marques E P, Wang H and Zhang J, 2007, *J Power Sources*,**173**, 891.

[13] Xiong L and Manthiram A, 2005, *J Electrochem. Soc.*, **152**, A697.

[14] Wang D L, Xin H L L, Hovden R, Wang H S, Yu Y C, Muller D A , DiSalvo F J and Abruna H D. 2013, *Nat. Mater.* **12**, 81.

[15] Valisi A N, Maiyalagan T, Khotseng L, Linkov V and Pasupathi S. 2012, *Electrocatalysis*, **3**, 108.

[16] Pryadchenko V V, Belenov S V, Shemet D, Volochaev V A, Srabionyan V V, Avakyan L A, Tabachkova N Yu, Guterman V E and Bugaev L A, 2017, *Journal of Technical Physics*, **59**, 1642. 10.21883/FTT.2017.08.44771.43.

[17] Chepkasov I V, *Physicochemical aspects of studying clusters, nanostructures and nanomaterials*, 2016, **8**, 387-392.

[18] Tsyganov S et al., *Physical Review B*, 2007, V 75, **4**, 045421.

[19] Jelinek B et al., *Physical Review B*, 2012, V 85 **24**, 245102.

[20] Nomoev A V, Yuzhapova N V, *Bulletin of the Russian Academy of Sciences. Physical series*, 2019, V 83 **11**, 1563-1567.– DOI 10.1134/S036767651911019X.

Changes in the immitance spectra of the cement paste in the initial period of hardening

V A Zhuravlev[1], D O Ugodenko[1], A V Zhuravlev[1], V I Suslyaev[1], I A Prischepa[2], Yu S Sarkisov[2], N P Gorlenko[2], A I Kudyakov[2]

[1] Department of Radiophysics, National Research Tomsk State University, 36 Lenin Ave., Tomsk, Russia
[2] Department of Physics, Chemistry and Theoretical Mechanics, Tomsk State University of Architecture and Building, 2 Solyanaya sq., Tomsk, Russia

ptica@mail.tsu.ru

Abstract. The paper analyzes changes in the impedance and admittance spectra of the cement slurry during hardening. The spectra were measured in the frequency range from 20 Hz to 2 MHz on a precision RLC meter Agilent E4890A using a homemade capacitor measuring cell. The studies in order to establish the optimal equivalent electrical circuits describing the transformation of the electrophysical properties of the cement slurry and interface phenomena in the near-electrode layer of the measuring cell from the time of hardening were carried out. The sensitivity of the parameters of equivalent circuits to different stages of cement slurry hardening has been established.

1. Introduction
Currently, a number of well-developed modern physical, physicochemical and chemical methods are widely used to determine the quality, structure and properties of various building materials. The use of impedance spectroscopy and dielectrometry methods along with them makes it possible to evaluate additionally a number of important characteristics of water-containing building materials, such as water-cement mixtures, mortars and concretes [1, 2].

The methods of the impedance spectroscopy and dielectrometry are very sensitive to the molecular mobility of water molecules in various states and ionic complexes dissolved in it. In turn, the structural state of water largely determines the most important properties of artificial stone, including mechanical strength [3]. Therefore, scanning the impedance characteristics of water-containing building materials in continuous time mode is of considerable interest. It can provide important information about the processes occurring both in the initial period of the hydration interaction of cement with water, and at the subsequent stages of setting and hardening of this complex dispersed system [4, 5].

The aim of this work is to study changes in the immittance spectra (impedance and admittance) of cement paste samples in an electrochemical measuring cell during the twenty-four hours, as well as to compile and analyze equivalent electrical circuits describing these changes.

2. Samples for research and method of measuring immittance spectra
Experimental studies were carried out on the samples of cement paste of normal density (ND = 26%) made of Portland cement CEM I 42.5N (GOST 30515–2013) of the Topkinsky cement plant and tap

IOP Conf. Series: Materials Science and Engineering 1198 (2021) 012014 doi:10.1088/1757-899X/1198/1/012014

water. The cement paste was prepared in a separate container. After thorough mixing, it was placed in the measuring cell, and the process of measuring its electrophysical characteristics began.

A precision RLC meter Keysight E4980A to measure the spectra of the electrophysical characteristics in the frequency range 20 Hz - 2 MHz at the initial time of the structure formation of the cement paste was used. The measuring cell was a flat capacitor consisting of two rectangular stainless steel plates with a thickness of 1 mm and dimensions 51.0 x 39.5 mm. The distance between the plates was 9.5 mm. The measurement results at intervals of 30 min during the day using a program code written in the LabVIEW environment were recorded.

According to an equivalent circuit consisting of a parallel connected capacitance (*C*) and a resistor (*R*) in this series of experiments the measurement results were recorded. The recalculation of the measured *R-C* parameters into the impedance $Z = Z' + iZ''$ and admittance $Y = Z^{-1} = Y' + iY''$ components was carried out according to the formulas given in Table 1.

Table 1. Formulas for converting the measured *R-C* parameters of the equivalent circuit into impedance, admittance components and vice versa.

	R-C parall.	$Z = Z' + iZ''$	$Y = Y' + iY''$
R-C parall.		$Z' = \frac{R}{1+(\omega CR)^2}$ $Z'' = \frac{-\omega CR^2}{1+(\omega CR)^2}$	$Y' = R^{-1}$ $Y'' = \omega C$
Z	$\omega C = \frac{-Z''}{Z'^2 + Z''^2}$ $R = \frac{Z'^2 + Z''^2}{Z'}$		$Y' = \frac{Z'}{Z'^2 + Z''^2}$ $Y'' = \frac{-Z''}{Z'^2 + Z''^2}$
Y	$\omega C = Y''$ $R = 1/Y'$	$Z' = \frac{Y'}{Y'^2 + Y''^2}$ $Z'' = \frac{-Y''}{Y'^2 + Y''^2}$	

3. Experimental results

3.1. Parallel R-C equivalent circuit

Figure 1 shows the frequency dependences of the electrical resistance and capacitance of a measuring cell filled with cement paste, measured at the initial ($t = 0$ min), final ($t = 1720$ min) moments of time and for the hardening time $t = 900$ min.

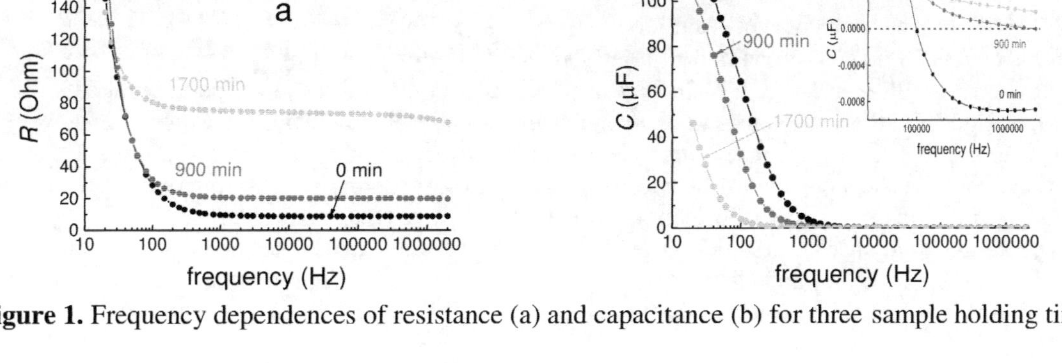

Figure 1. Frequency dependences of resistance (a) and capacitance (b) for three sample holding times.

According to Fig. 1a, the resistance values R of the sample rapidly decreases at frequencies up to ~ 100 Hz, and then change slightly with increasing frequency. In the frequency range above 100 Hz, an increase in resistance is observed with increasing holding time. Figure 1b shows that a rapid decrease in capacitance with increasing frequency and increasing exposure time also occurs at low frequencies up to ~ 1000 Hz.

In connection with the above, the analysis of the time dependences of the capacitance and resistance values for a measurement frequency of 200 Hz was carried out. The results of measurements are shown in Figure 2.

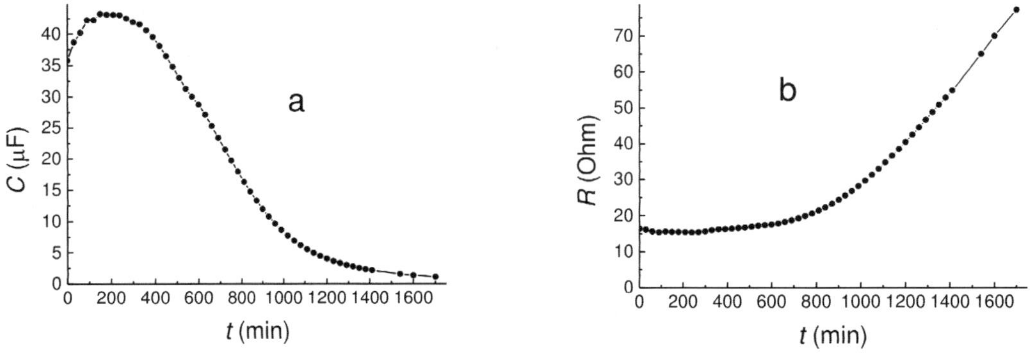

Figure 2. Dynamics of changes in capacitance (a) and resistance (b) in the "cement-water" system at the initial time of hardening. The measurements were carried out at a frequency of 200 Hz.

According to Fig. 2a, from the moment of mixing the cement with water, the sample capacitance is first to increase up to a time of ~ 200 min. After passing the maximum $C \approx 43$ μF, at first a rapid, and at $t >$ 900 min, a slower decrease in the capacitance to the value $C \approx 1$ μF at $t = 1720$ min occurs. Figure 5b shows that the resistance of the sample changes slightly up to the holding time $t \approx 700$ min. Further, there is a more rapid increase in the R values. Thus, the largest changes in the capacitive component occur at the initial stage of cement paste hardening. Whereas the main changes in the resistive component, occur at the final stage.

The high-frequency section of the $C(f)$ dependence for frequencies $f \geq 100$ kHz is shown in the inset to Figure 1b. The horizontal dashed straight line in the inset corresponds to zero capacitance. It can be seen that at exposure times from 0 to 900 min the capacitive component has a negative sign in the investigated frequency range. This indicates a change in the character of the sample impedance from capacitive to inductive. Thus, the equivalent circuit consisting of a parallel connected resistor and capacitor is not suitable for describing the behavior of the electrophysical characteristics of the "cement-water" system in these frequency-time intervals.

3.2. Analysis of immitance spectra of cement paste

In this section, we will focus on the choice of the parameters of the equivalent circuit that describe the time evolution of the immitance spectra of the cement paste during its hardening. For this, the measured parameters of the equivalent R-C circuit into impedance $Z = Z' + iZ''$ and admittance $Y = Z^{-1} = Y' + iY''$ components according to the formulas in Table 1 were converted.

The spectra $Z(f)$ and $Y(f)$ obtained at hourly intervals are presented as hodographs of impedance and admittance in Figures 3 and 4, respectively. The points in the figures are experimental data. Lines are calculations for the parameters of the equivalent circuits obtained below. Figure 3a shows the measurement results over the entire range of exposure times from $t = 0$ min to $t = 1720$ min. Figure 3b shows on a larger scale a portion of the spectra up to the exposure time $t = 900$ min, in which a change in the sign of the imaginary part of the impedance at high frequencies is observed.

The low-frequency parts of the hodographs shown in Fig. 3 are typical for non-blocking electrodes with double electric layers and diffusion processes occurring near the interface. Moreover, electric double layers are formed also at the cement - electrolyte phase boundaries. In the initial period of hardening of the cement paste, the angle of inclination of the low-frequency branches of the hodographs increases with time. Moreover, after a holding time of $t \approx 300$ min, the tilt angle practically does not change. A constant phase element (CPE) is well suited for describing such impedance spectra [6 - 8]. This element was proposed first for the analysis of electrochemical processes and dielectric relaxation processes in [9]:

$$Z_{CPE} = A(i\omega)^{-S} = A\omega^{-S}\,[\cos(\pi S/2) - i\sin(\pi S/2)]. \quad (1)$$

Here $i = \sqrt{-1}$, A and S are phenomenological parameters describing interactions in the near-surface electrode-liquid layer and at boundaries of the phase, as well as interlayer polarization processes. If $S = 0$, then $Z = A = R$, that is, the impedance is resistive. If $S = 1$ then $Z = 1/(i\omega A^{-1})$, that is, the impedance is capacitive ($A^{-1} = C$). At $S = -1$, the impedance is inductive $Z = i\omega A$ ($A = L$).

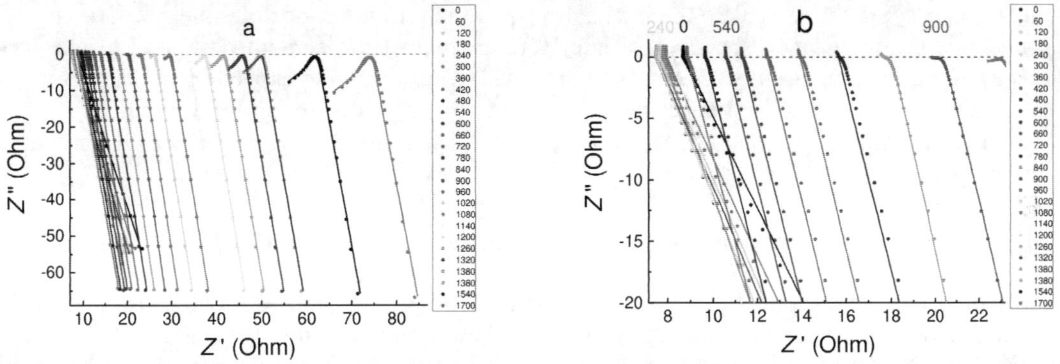

Figure 3. Impedance hodographs over the entire range of cement paste holding times (a) and over the time interval of existence of a change in the sign of Z'' (b). The numbers in Fig. 3b are the measurement times of some spectra from the moment of mixing. Points are experiments, lines are calculations.

The point of intersection of the hodographs with the abscissa at $Z'' = 0$ determines the value of the active resistance R, which is the sum of two quantities. One of them characterizes ohmic losses in the interface region of the sample - electrode. The second describes ohmic losses in the sample. Note that at holding times greater than 900 min, the inductive impedance component ($Z'' \geq 0$) disappears and another high-frequency dispersion region appears in the impedance spectra. It is associated with the processes of dielectric polarization in the emerging particles of cement stone.

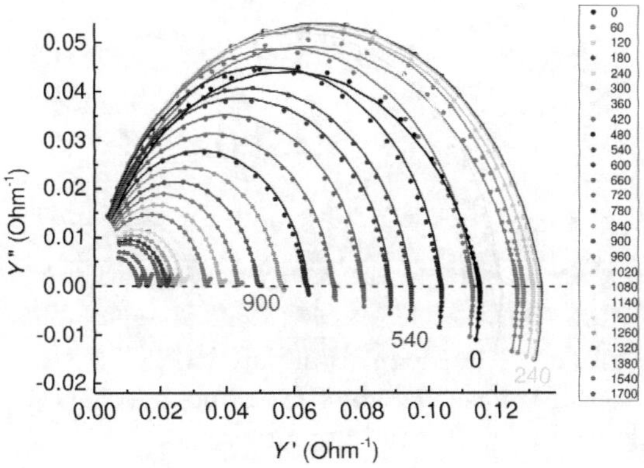

Figure 4. Admittance hodographs. The numbers in the figure are the measurement time of some spectra from the moment of mixing. Points are experiments, lines are calculations.

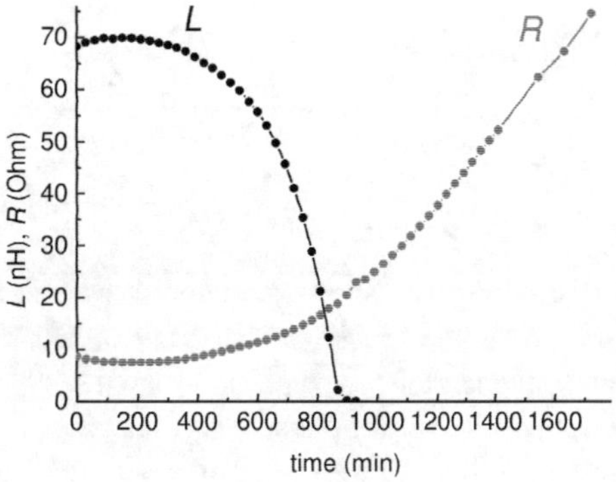

Figure 5. Estimates of the values of inductance (L) and resistance (R) depending on the holding time of the cement paste.

IOP Conf. Series: Materials Science and Engineering 1198 (2021) 012014 doi:10.1088/1757-899X/1198/1/012014

In order to characterize in more detail the electrochemical processes occurring at the initial moment of cement paste hardening, we supplemented the analysis of impedance spectra with calculations of admittance spectra. These spectra are shown in Figure 4. It can be seen that they provide more visual information about the immitance spectra at the initial stage of cement paste setting. Whereas the impedance plots in Figure 3a describe the final stage in more detail.

Based on the above considerations, we used two equivalent circuits to describe theoretically the immittance spectra of the cement paste at different holding times. The first circuit contains elements connected in series: inductance L, resistor R and constant phase element Z_{CPE1}:

$$Z_1 = R + Z_L + Z_{CPE1} = R + i\omega L + A_1(i\omega)^{-S_1}. \tag{2}$$

Formula (2) was applied to calculate the impedance spectra up to the exposure time $t = 900$ min.

To describe the impedance spectra at a later setting time of the cement paste, we tested various versions of the equivalent circuits. They were composed of resistors and capacitors connected in series and in parallel. However, the best agreement with experiment was obtained for a circuit in which the emerging high-frequency dispersion region is described by another CPE connected in parallel with resistor R:

$$Z_2 = [RZ_{CPE2}]/[R + Z_{CPE2}] + Z_{CPE1} = [RA_2(i\omega)^{-S_2}]/[R + A_2(i\omega)^{-S_2}] + A_1(i\omega)^{-S_1}. \tag{3}$$

The parameters included in the equivalent circuits (2) and (3) were estimated using a program written in the VBA programming language in the Excel environment using the Solver add-in. Figure 5 shows the estimated values of inductance (L) and direct current resistance (R) depending on the setting time of the cement-water mixture. The time dependences of these parameters are similar to the time dependences of capacitance and resistance at a frequency of 200 Hz presented in Figure 2. Only the value of the inductance decreases with time much faster than the capacitance and turns to zero at t = 900 min. In addition, the value of the resistance at direct current in the initial hardening period of the cement-water mixture is approximately two times less than at a frequency of 200 Hz.

The estimates of the parameters characterizing the magnitude (A_i) and exponent (S_i) of the impedances of the constant phase elements Z_{CPE1} and Z_{CPE2} are presented in Figures 6 and 7, respectively.

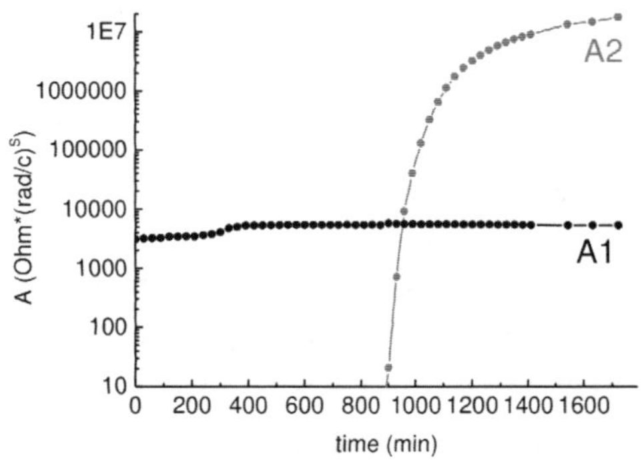

Figure 6. Dependences of parameters A_1, A_2 of elements with constant phase Z_{CPE1} and Z_{CPE2} depending on the holding time of the cement paste.

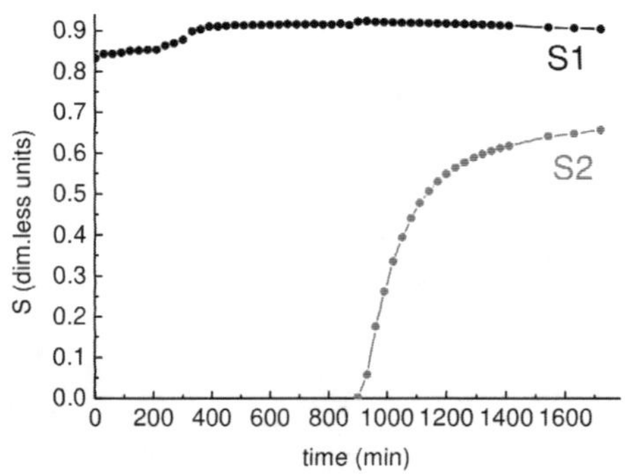

Figure 7. Dependences of parameters S_1, S_2 of constant phase elements Z_{CPE1} and Z_{CPE2} depending on the holding time.

From Figures 6, 7 it can be seen that the properties of the Z_{CPE1} element change in the entire range of variation of the holding time not as significantly as the parameters of the Z_{CPE2} element. Thus, $A_1(t=0)$ is equal to 3100 Ohm(rad/c)$^{S}{}_1$, and $A_1(t=1720)$ is equal to 5500 Ohm(rad/c)$^{S}{}_1$. The value of the exponent S_1 is close to one. This indicates that the immitance of this CPE is close to the capacitive immitance. A noticeable break in the dependences $A_1(t)$ and $S_1(t)$ is observed in the vicinity of the holding time $t \approx 300$ min, when the quantities $C(t)$ and $L(t)$ pass through a maximum.

The parameters of the second CPE, describing the appearance of a new region of dispersion after 900 minutes of holding the cement paste, increase sharply. Then they change more smoothly and their time dependences tend to saturation.

4. Discussion

The processes of dissolution, dispersion, hydration and ion exchange are at the heart of the phenomena that occur after the mixing of cement with water. Later, the processes of adsorption and enlargement of particles supplement them during the crystallization of cement stone. Without dwelling on these processes in detail due to the limited space, we note the following.

After mixing the cement with water, the processes of the initial hardening of the cement paste begin after about 90 minutes of exposure and continue up to about 240 minutes. According to Figures 2a and 5, at this time the values of capacitance $C(t)$ and inductance $L(t)$ reach a maximum and then begin to decrease rapidly. The value of $R(t)$ varies little over this time interval, and kinks are observed in the dependences of the parameters of Z_{CPE1}.

The end-of-setting time for Portland cement is approximately 720 minutes. A sharp transition from the plastic strength of the cement paste to the brittle strength of the hardened cement stone occurs after 720 – 840 minutes of hardening. After 840 – 900 minutes, a period of intense crystallization begins. In this case, gel-like neoplasm's approach each other and form crystals. Then the crystals grow together and form spatial structures. At this stage of the maturation of the cement paste, the inductive component of the immitance vanishes and a high-frequency dispersion region appears, described by the Z_{CPE2} element. Obviously, it is associated with polarization processes inside the particles of the formed cement stone.

Equivalent circuit element Z_{CPE1} describes the low-frequency polarization region. Responsible for this area are the processes of polarization in the near-electrode double electric layer, in the double electric layers at the phase boundaries: "cement grain - electrolyte" and the processes of interlayer polarization at the grain boundaries. During the hardening of the cement paste, the specific contribution from these processes changes, since the amount of free water decreases, and the amount of bound water increases. However, the redistribution of contributions occurs in such a way that the parameters of the resulting process do not change significantly.

5. Conclusion

Thus, the conducted studies of the immittance spectra of the cement paste made it possible to determine the type of equivalent electrical circuits describing the transformation of the spectra at different stages of hardening. The correlation of time dependences of parameters of equivalent circuits with different stages of maturation of cement paste were established.

Acknowledgments
Tomsk State University Competitiveness Improvement Program supported research.

References

[1] Lianzhen X and Xiastes W 2011 *Journal of Wuhan University of Technology — Materials Science Edition.* **22** 983
[2] Heikal M, Helmy I, El-Didamony H and El-Raoof F 2004 *Ceramics-Silikaty* **48** 49
[3] Salem Th M 2002 *Cement and Concrete Research* **32** 1473
[4] Liu K, Cheng X, Li J, Gao X, Cao Y, Guo X, Zhuang J and Zhang C 2019 *Composites Part B* **177** 107435
[5] Yoon S S, Kim H C and Hill R M 1996 *Journal of Physics D: Applied Physics* **29** 869
[6] Macdonald J R 1984 *Solid State Ionics* **13** 147
[7] Barbero U, Becchi M and Freire F C M 2008 *J. Appl. Phys.* **104** 114111
[8] Zhuravlev V A, Suslyaev V I, Zhuravlev A V and Korovin E Yu 2018 *Russ. Phys. J.* **60** 1893
[9] Cole K S and Cole R H 1941 *J. Chem. Phys.* **9** 341

IOP Conf. Series: Materials Science and Engineering 1198 (2021) 012015 doi:10.1088/1757-899X/1198/1/012015

Synthesis of textured films of hexagonal barium ferrite on Al_2O_3/Si_3N_4 amorphous structure

A. Yu. Mironovich[1], V.G. Kostishyn[1] and I.M. Isaev[1]

[1] National University of Science & Technology «MISiS», Moscow, Russia

E-mail: amironovich24@gmail.com

Abstract. In this work structural studies of $BaFe_{12}O_{19}/Al_2O_3/Si_3N_4$ thin films with different thicknesses of aluminum oxide and silicon nitride are represented. It is shown that barium hexaferrite crystallizes on the amorphous surface of aluminum oxide with spontaneously formed (00l) uniaxial texture. Microstructural differences in $BaFe_{12}O_{19}$ films were observed with varying of the thickness of amorphous sublayers, which is explained by the effect of mechanical stress.

1. Introduction

$BaFe_{12}O_{19}$ hexagonal barium ferrite (BaM) is well-known essential material in microwave technology and radio electronics. At present, an intensive search and study of materials based on BaM are underway for new generation microwave devices, in particular, integrated with semiconductor planar technology. Technologies for producing anisotropic BaM films are being actively developed in this area. However, due to the peculiarities of the material, the synthesis of such films is associated with certain difficulties [1].

Usually, to obtain anisotropic films, the principle of epitaxial growth is used and single-crystal substrates are required. In this context, the possibility of obtaining such films on amorphous substrates is advantageous and eliminates the need for single crystals. Spontaneous textured growth is based on the principle of selective formation of film nuclei with an orientation that provides a minimum of surface and interfacial energies [2]. In materials with a close-packed structure, energy is minimized by orienting the packing direction perpendicular to the film surface. Thus, the (111) texture is observed for metals with a face-centered cubic unit cell, and for materials with a hexagonal close packing, such as hexaferrites, the texture type is (001).

After deposition, the films do not have sufficient crystallinity and, as a consequence, have no magnetic ordering. For the formation of the crystalline barium hexaferrite from such an amorphous precursor, annealing at temperatures above 800 °C is required. This raises a problem of a different kind. During annealing, the material crystallizes throughout the entire volume of the film, i.e. in energetically nonequivalent positions: on the surface, at the film-substrate interface, and in the bulk. In these positions, the most favorable crystallite orientation may differ. Thus, near the film / substrate interface, crystallites are formed with an orientation inherited from the substrate. On the surface grains with minimal surface energy crystallize. In the volume, the orientation of the grains can either be random or determined by the minimization of mechanical stresses. In the case of hexaferrite on an amorphous substrate, the largest contribution to the deterioration of the (00l) texture is expected from bulk crystallites. However, their formation can be avoided. As shown in [3, 4], in hexaferrite films

IOP Conf. Series: Materials Science and Engineering 1198 (2021) 012015 doi:10.1088/1757-899X/1198/1/012015

with thickness less than 100 nm crystallization of randomly oriented grains does not occur. Thus, to avoid crystallization of non-oriented grains in the bulk of the film, in this work, the hexaferrite thickness also did not exceed 100 nm.

Obtaining a textured $BaFe_{12}O_{19}$ film on amorphous substrates, texture perfection assessment and investigation of the effect of the thickness of amorphous underlayers are the objectives of this work.

2. Experimental

Silicon nitride films were synthesized on single-crystal silicon wafers by plasma enhanced chemical vapor deposition method (Corial D250) with the following process parameters: substrate temperature 250 °C, operating pressure 2 Torr, RF generator power 90 W, gas flow rate 50, 180, 1500, and 100 sccm for SiH_4, NH_4, N_2, and Ar, respectively. Films of aluminum oxide and barium hexaferrite were deposited by ion-beam sputtering. A polycrystalline plate of Al_2O_3 and a $BaFe_{12}O_{19}$ disk prepared by standard ceramic technology were used as targets. The deposition was proceeded in a UVN-71 apparatus with a following parameters: the discharge current of the ion source was 40 mA, the discharge voltage was 2 kV, the working gas (Ar) pressure was 0.3 mTorr, and the substrate temperature was 300–330 °C. After deposition, the BaM films were annealed for 1 hour at 900 °C with 5 °/min heating rate and natural cooling.

The thickness of the films was estimated using a contact profilometer (DekTak 150). For all samples, the BaM thickness was approximately 100 nm. The set of thicknesses for Al_2O_3 and Si_3N_4 is presented in Table 1.

X-ray diffraction (XRD) patterns were obtained on a BRUKER D8 ADVANCE diffractometer (Cu Kα, λ=0.154 nm, U=40 kV, I=40 mA). The surface morphology of the films was studied by atomic force microscopy (AFM, NT-MDT NTEGRA Prima).

Table 1. Al_2O_3/Si_3N_4 thickness ratio

Sample, №	Al_2O_3 thickness, nm	Si_3N_4 thickness, nm
1	200	50
2	200	100
3	200	200
4	100	100
5	50	100

3. Results and discussion

The XRD patterns of the obtained samples do not fundamentally differ from each other and show only the peaks of the BaM (00l) and of the silicon substrate. The only difference is in the ratio of the intensities of reflexes of different order of reflection, which is not essential in the context of this work. Figure 1 demonstrates a typical X-ray pattern of synthesized sample. Thus, X-ray phase analysis confirms the formation of a textured BaM film and amorphous state of underlayers. On the other hand, there is evidence according to which the film surface could be covered with grains of various orientations, which are undetectable by XRD, but significantly affect the anisotropy of the magnetic properties [5].

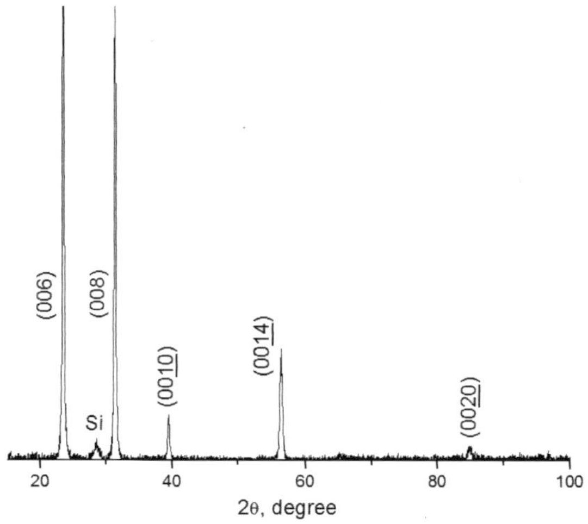

Figure 1. X-ray diffraction pattern of sample № 4

Nevertheless, AFM measurements showed the absence of non-oriented grains on the films surface. All crystallites have a circular shape, which is characteristic of the (00l) texture [3, 4]. The ratio of the Al_2O_3/Si_3N_4 thicknesses showed the influence on the microstructure of the BaM films. In particular, Figure 2 shows the AFM images of the surface of samples № 1 and № 2. It can be seen that in film № 1 the relief inhomogeneity is small, and the grains are densely adjacent to each other. In the sample № 2, the observed height difference is more noticeable, and voids are clearly visible between the crystallites. Such microstructural differences are caused by the mechanical stresses in the film, the magnitude of which depends on the thickness of the film and its material [6].

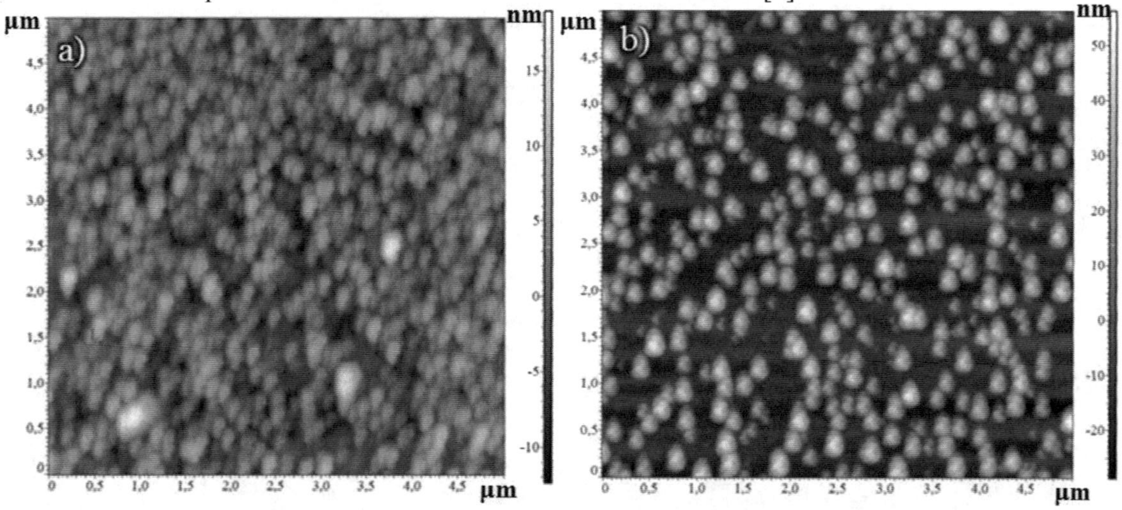

Figure 2. AFM images of samples №1 (a) and №2 (b)

In addition to microstructural differences, the magnitude of mechanical stresses affects the presence of macroscopic defects that destruct the film. In Si_3N_4 and Al_2O_3 films, the stresses have opposite signs and can be compensated [7]. Of all the samples synthesized, sufficient stress compensation was observed for sample №2. As can be seen (Figure 3), there are no macroscopic defects on the surface of this film, while the rest of the samples exhibit blistering of various radii and concentrations. In addition, the size and number of blistering increased over time, to the point that they became visible to the naked eye. Thus, only sample №2 can be suitable for practical applications. On the other hand, its microstructure is less favorable for such applications as magnetic memory devices, for which grain uniformity is important. However, microstructure can be changed by adjusting the temperature of the substrate during film deposition [8], so the obtained material still has a certain potential for practical application.

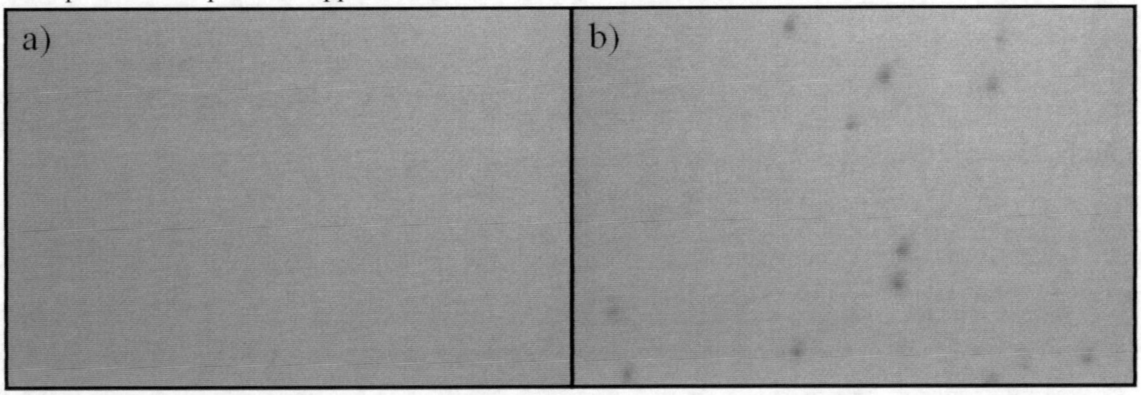

Figure 3. Optical images of the samples №2 (a) and №4 (b). The scale is equal, the magnification is x20

4. Conclusions

Textured thin films of $BaFe_{12}O_{19}/Al_2O_3/Si_3N_4$ with different thicknesses of silicon nitride and aluminum oxide were obtained. The high degree of crystallographic texture of the BaM film is confirmed by X-ray diffraction and atomic force microscopy. In such films the crystallographic c axis of hexaferrite is directed perpendicular to the plane of the film in all grains. As the [001] direction is also the easy magnetization axis, it will determine the anisotropy of the magnetic properties. It is shown that the microstructure changes with varying the thickness of Al_2O_3 and Si_3N_4. This fact is important since the magnetic parameters (especially the coercive force) depend on the microstructure.

References

[1] Kostishin V G, Mironovich A Yu, Shakirzyanov R I, Isaev I M and Sergienko A A 2020 *Usp. Prikl. Fiz.* **8** 370
[2] Glezer A M and Shurygina N A 2013 *Amorphous-nanocrystalline alloys* (Moscow: Fizmatlit)
[3] Cho T S, Doh S J, Je J H and Noh D Y 1999 *J. Appl. Phys*. **86** 1958
[4] Xu Z, Lan Z, Sun K, Guo R, Yu Z, Jiang X and Zhu G 2013 // *Appl. Surf. Sci.* **271** 362
[5] Liu X H, Hong M H, Song W D, Chen G X, Chong J F, Wang J P, Wu Y H and Chong T C 2004 *Appl. Phys.* A. **78** 423
[6] Abadias G., Chason E.,Keckes J, Sebastiani M, Thompson G B, Barthel E, Doll G L, Murray C E, Stoessel C H snd Martinu L 2018 *J. Vac. Sci. Technol.* A. **36** 20801
[7] Kostishin V G, Mironovich A Yu, Timofeev A V, Shakirzyanov R I, Isaev I M, kurochka A S, Ril A I and Sergienko A A 2021 *Russ. J. Inorg. Chem.* **66** 603.
[8] Kostishin V G, Mironovich A Y, Shakirzyanov R I, Isaev I M, Timofeev A V, Ril A I and Lega P V *2021 J. Magn. Magn. Mater.* **527** 167786

AUTHOR INDEX

Institute of Physics
Dirac House, Temple Back
Bristol BS1 6BE UK

ISSN: 1757-8981
ISBN 978-1-7138-4657-4

CURRAN ASSOCIATES INC.
proceedings.com